1st Asian ICMC & CSSJ 50th Anniversary Conference (1st Asian ICMC-CSSJ 50)

Journal of Physics: Conference Series Volume 897

Kanazawa, Japan
7 - 10 November 2016

Editors:

Mitsuho Furuse
Takanobu Kiss
Jun-ichi Shimoyama

ISBN: 978-1-5108-4867-2
ISSN: 1742-6588

Printed from e-media with permission by:

Curran Associates, Inc.
57 Morehouse Lane
Red Hook, NY 12571

Some format issues inherent in the e-media version may also appear in this print version.

Printed by Curran Associates, Inc. (2017)

For permission requests, please contact the Institute of Physics
at the address below.

Institute of Physics
Dirac House, Temple Back
Bristol BS1 6BE UK

Phone: 44 1 17 929 7481
Fax: 44 1 17 920 0979

techtracking@iop.org

Additional copies of this publication are available from:

Curran Associates, Inc.
57 Morehouse Lane
Red Hook, NY 12571 USA
Phone: 845-758-0400
Fax: 845-758-2633
Email: curran@proceedings.com
Web: www.proceedings.com

Table of contents

Volume 897

1st Asian ICMC & CSSJ 50th Anniversary Conference
7–10 November 2016, Kanazawa, Japan

Accepted papers received: 11 September 2017
Published online: 21 September 2017

Preface

Cryogenic Materials

Cryocooler and Cooling Systems

Magnet, Power Device and Systems

IOP Conf. Series: Journal of Physics: Conf. Series **897** (2017) 011001 doi:10.1088/1742-6596/897/1/011001

Preface

It is our great pleasure to report that the Proceedings of joint meeting of the 1st Asian ICMC CSSJ 50th Anniversary Conference (1st Asian ICMC-CSSJ 50) held at Kanazawa Kagekiza, Japan from November 7th to 10th, 2016 is published in *Journal of Physics: Conference Series* (JPCS). The 1st Asian ICMC-CSSJ 50 organized by the International Cryogenic Materials Conference in Asia (Asian ICMC) and Cryogenics and Superconductivity Society of Japan (CSSJ), was aimed to gather a large number of scientists, engineers, students, corporate executives and other participants from Asian countries, and to promote technologies on superconductivity and cryogenics through presentations and fruitful discussions. Motivations of this new international conference are to enhance communications between researchers in this field in Asia and to celebrate 50 years anniversary of CSSJ established in 1966. Total numbers of participants and presentations of the 1st Asian ICMC-CSSJ 50 are 378 from 16 countries and 264 including 5 plenary lectures, 5 invited talks for Asian topics and 64 invited talks, respectively. This conference consisted of many sessions, such as Superconducting Materials, Critical Current Properties of Fe-Based Superconductors, Wires and Tapes, Evaluation and Analysis, Stability, AC Losses, FCLs/Transformers, Power Applications, Cables, Magnet Technology, Magnet Design, MRIs, Fusion, SMESs / Combination with RE, LTS Strands, Superconducting Joints, Rotating Machines, Superconducting Bulks, Fabrication and Application of Superconducting Bulk, Magnetization of Superconducting Bulk, Magnetic Separation, LTSs/Applications of Cryogenics, Large Scale Cryogenics, Cryogenic Materials, Cryogenic Cooling Engineering, Cryocoolers, Heat Transfer, Material Properties for Cooling Devices, Cryogenic Structural and Insulation Materials, Electronics, SQUIDs and Detectors, Thin Films and Devices and Low Carbon Society.

To ensure the high publication standard mandated by JPCS, every paper was peer reviewed by a reviewer with expertise, and in some cases by two reviewers before it was accepted for publication. The reviewing process was managed by the program committee of CSSJ. As editors of the Proceedings, we would like to express our sincere appreciation to all the reviewers involved in the evaluation of the papers for their invaluable contribution.

Editors and Program Committee are listed below.

Editors: Mitsuho Furuse (AIST), Takanobu Kiss (Kyusgu Univ.), Jun-ichi Shimoyama (Aoyama Gakuin Univ.)

IOP Conf. Series: Journal of Physics: Conf. Series **897** (2017) 011001 doi:10.1088/1742-6596/897/1/011001

Program Committee: Nobuya Banno (NIMS), Hiroyuki Fujishiro (Iwate Univ.), Mitsuho Furuse (AIST), Mutsuo Hidaka (AIST), Tsutomu Hoshino (Meisei Univ.), Yota Ichiki (Hitachi), Masayoshi Inoue (Kyushu Univ.), Masataka Iwakuma (Kyushu Univ.), Akifumi Kawagoe (Kagoshima University), Hiroshi Miyazaki (Toshiba), Takahiro Okamura (KEK), Takeshi Shimazaki (AIST), Jun-ichi Shimoyama (Aoyama Gakuin Univ.), Suguru Takada (NIFS), Yutaka Yamada (Shibaura Institute of Technol), Tetsumi Yuri (NIMS)

IOP Conf. Series: Journal of Physics: Conf. Series **897** (2017) 011002 doi:10.1088/1742-6596/897/1/011002

Peer review statement

All papers published in this volume of *Journal of Physics: Conference Series* have been peer reviewed through processes administered by the proceedings Editors. Reviews were conducted by expert referees to the professional and scientific standards expected of a proceedings journal published by IOP Publishing.

IOP Conf. Series: Journal of Physics: Conf. Series **897** (2017) 012001 doi:10.1088/1742-6596/897/1/012001

Development of point-contact spectrometer for spin polarization measurements

M. Shiga[1], N. Nishimura[1], H. Takata[1], Y. Inagaki[1], H. Kambara[2], K. Tenya[2], and T. Kawae[1]
[1]Department of Applied Quantum Physics, Faculty of Engineering, Kyushu University, Motooka, Fukuoka 819-0395, Japan

[2]Faculty of Education, Shinshu University, Nagano, 380-8544, Japan

e-mail 2TE15295Y@s.kyushu-u.ac.jp

Abstract. We construct a "needle-anvil" type point-contact spectrometer for spin polarization measurements, which is installed in a ^{4}He-cryostat. Two types of piezo devices are used to control the contact size precisely between a sample ferromagnet and a superconductor. An attocube piezo-based positioner is mounted for coarse movement of the tip, while a stacked-type piezo device is used for fine control of the contact size. This enables to change the contact size between the tip and sample from sub micro-meters to atomic-size contacts continuously. By suppressing thermal flow into the sample space and mechanical vibration, we can keep the contact over hours, enabling the precision measurements. To examine the performance of the spectrometer, we study the spin polarization of polycrystalline $SrRuO_3$ with the point-contact Andreev reflection measurements. The polarization is estimated to be ~0.59 at the clean limit of the interface, which is consistent with previous study.

1. Introduction

Point-contact spectroscopy (PCS) is used to study the interaction of conduction electrons with excitations of quasiparticles [1]. When the contact size d is smaller than electron mean free path l, the conduction electron is accelerated by the bias voltage and moved without energy dissipation in the contact, which is known as "ballistic transport". Then, the ballistic electrons lose their energy by interacting with a variety of excitations such as phonons. When the ballistic electron enters from metal to superconductor, the characteristic features is expected to appear at superconductor/metal interface depending on the energy of ballistic electron. In the region that the energy of incident electron is lower than the superconductor energy gap Δ, the electron can form the cooper pair at the interface as shown in figure 1. Simultaneously, the hole reflects from superconductor to metal. This phenomenon is called as Andreev reflection.

At the superconductor/ferromagnet interface, the Andreev reflection is suppressed due to the difference of density of state (DOS) between the up spin and down spin at the fermi surface. This difference of the DOS resulting in the spin polarization expressed by the following equation,

$$P = \frac{D_{Ef}^{\uparrow} - D_{Ef}^{\downarrow}}{D_{Ef}^{\uparrow} + D_{Ef}^{\downarrow}} \qquad (1),$$

where $D^{\uparrow}_{Ef}$ and $D^{\downarrow}_{Ef}$ are the density of state of up spin and down spin at the fermi surface. This implies that the spin polarization in ferromagnetic materials can be investigated through Andreev reflection between the ferromegnet and superconductor [2]. By using the point-contact Andreev reflection (PCAR) measurements, the spin polarization P and superconducting energy gap Δ are demonstrated to be estimated by fitting the differential conductance (dI/dV) spectra between superconductor and ferromagnetic material to a modified Blonder-Tinkham-Klapwijk (mBTK) model [3, 4]. Figure 1(b) shows P dependence of the conductance calculated from mBTK model, which is largely changed with the magnitude of P.

In this study, we develop a "needle-anvil" type point-contact spectrometer for PCAR measurements, which is mounted in a ^{4}He cryostat. By using two types of piezo devices, the contact size of the superconductor/ferromagnet interface can be controlled very precisely. We have measured the spin polarization of ferromagnet iron and polycrystalline $SrRuO_3$, which exhibits a ferromagnetic transition at $T \sim 160$ K, with this spectrometer. The present results provide high resolution compared to previous works.

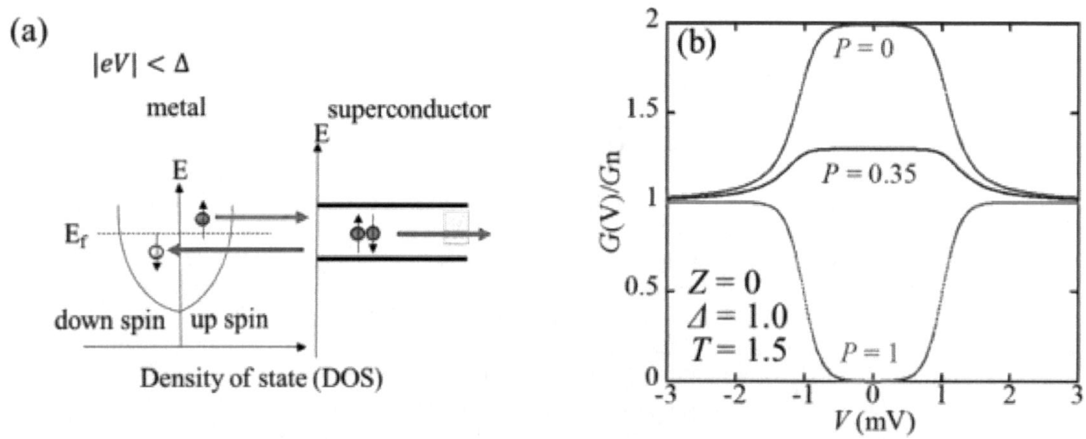

Figure 1. (a) The schematic diagram of Andreev reflection. The blue circle and white circle are electron and hole, respectively. In the superconductor, two electrons form a Coopet-pair (b) P dependence of the conductance at a superconductor/ferromagnet interface estimated by m-BTK model.

2. Experimental setup

Figure 2(a) shows a schematic illustration of the spectrometer. To control the contact size of the interface accurately, we use two types of piezo devices. An attocube piezo-based positioner (ANPz51 Attocube systems AG) is mounted for coarse movement of the tip, while a stacked-type piezo device is used to control the contact size. A superconducting tip is positioned to be ~3 mm apart from a sample at room temperature as depicted in left panel of figure 2(a). After cooling the temperature to $T \sim 4$ K, the tip is approached to contact softly with the sample by using the attocube positioner. Then, the contact size between the tip and sample is changed finely by the stacked-type piezo device as shown in the right panel of figure 2(a). Owing to this method, we can perform the PCAR measurements with changing the contact size without breaking the contact during the measurements, resulting in the smooth variation of interface scattering parameter. Figure 2 (b) shows a schematic diagram of cryostat (left panel) and an experimental setup (right panel). Note that thermal flow into the sample space is anchored at each thermal stage to keep the contact with the same size over a few hours.

The polycrystalline sample of $SrRuO_3$ is synthesized by the conventional solid-state reaction method. The powders of $SrCO_3$ and RuO_2 with stoichiometric compositions were carefully mixed and calcined

IOP Conf. Series: Journal of Physics: Conf. Series **897** (2017) 012001 doi:10.1088/1742-6596/897/1/012001

in air at 1100°C for 24 hours. After careful mixing of the calcined samples, they were shaped into pellets and then sintered at 1100 °C for 24 hours [5].

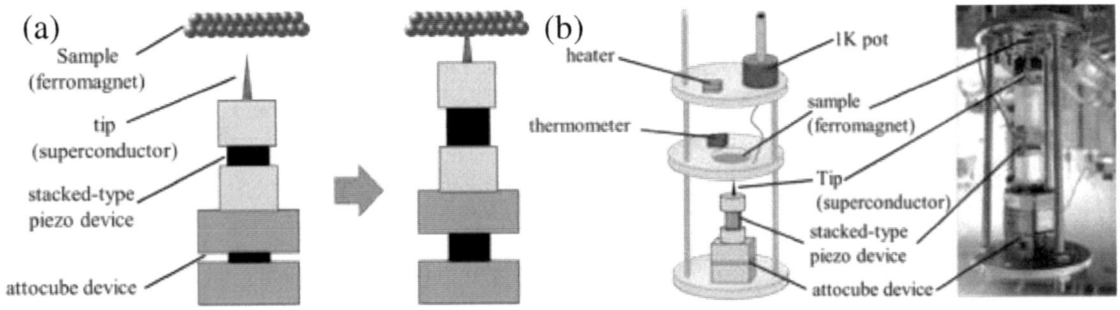

Figure 2. (a) The schematic diagram of a needle-anvil spectrometer for point-contact spectroscopy. (b) The schematic diagram and photo of experimental setup.

The conductance versus voltage (*G-V*) characteristics between a superconductor/ferromagnet interface are measured at $T = 4.2$ K in vacuum with a conventional lock-in technique with modulation frequency of 2 kHz. The superconducting Nb tip is fabricated by mechanically polishing a wire with the diameter of 0.2 mm. The sample is mounted in ^{4}He cryostat after polishing its surface. The spin polarization *P*, interface scattering parameter Z and superconducting energy gap Δ are obtained from fitting the differential conductance d*I*/d*V* to the m-BTK function.

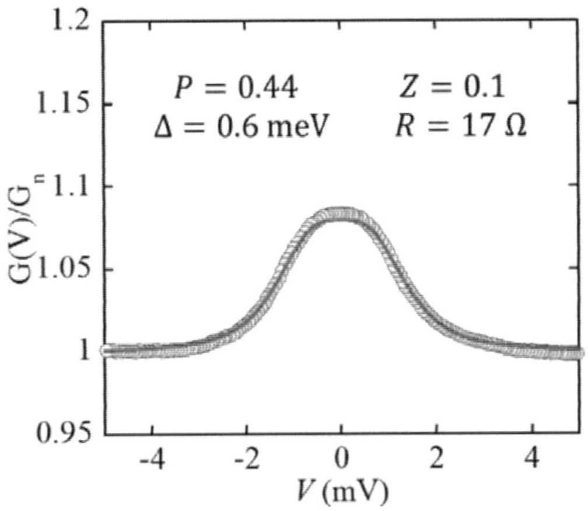

Figure 3. The differential conductance of Nb/Fe interface at $T = 4.2$ K. The circle and solid line are experimental data and fitting by m-BTK model. The fitting parameters are given in the figure.

3. Results and discussion

Figure 3 shows the differential conductance (d*I*/d*V*) spectra of a Nb/Fe interface, which is normalized at normal state conductance at the bias voltage $V = 5$mV. The experimental data are well reproduced by the m-BTK model. From the fitting, the spin polarization of Fe is estimated to be $P\sim0.44$, which is in good agreement with the previous study $P\sim0.43\pm0.03$ [6]. These results indicate that we can measure the spin polarization of ferromagnet precisely with using the new spectrometer. Note the superconducting energy gap of Nb is obtained to be $\Delta\sim0.6$ meV, which is smaller than that of bulk Nb with $\Delta_{\rm Nb} = 1.5$ meV. The suppression of the energy gap is probably caused by the pair-breaking effect by ferromagnetic iron.

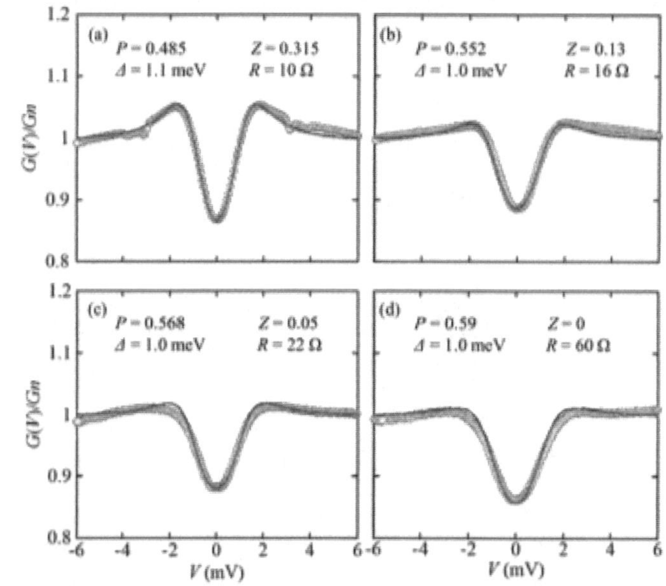

Figure 4. The representative spectra of differential conductance at $T = 4.2$ K. The spectra are measured with increasing the resistance at the interface without breaking the contact after preparing it. The solid line shows fitting by mBTK model.

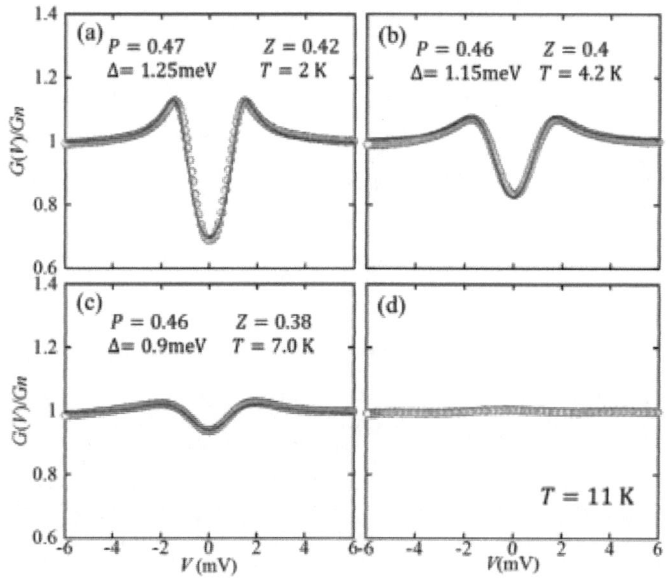

Figure 5. The temperature dependence of differential conductance of $SrRuO_3$/Nb interface from 2.0 K to 11 K. The circle and solid line are experimental data and fitting by m-BTK model.

Next we show the temperature dependence of the differential conductance spectra at a Nb/$SrRuO_3$ interface. The representative spectra at $T = 4.2$ K for various resistances are plotted in figure 4(a)-4(d), where the fitting curve by the m-BTK function and the fitting parameters are also shown. Note that the measurements are performed without breaking the contact after preparing it, leading to a change of the differential conductance curve. The superconducting gap of bulk Nb is 1.5 meV, which is larger than Δ ~1.0 mV estimated from the present experiments. As in the case of iron in figure 3, the gap is considered to be suppressed by magnetic scattering at the superconductor/ferromagnet interface. From these results, the spin polarization of polycrystalline $SrRuO_3$ is evaluated to be $P \sim 0.59$ [7]. This value is larger than $P_1 = 0.51$ in a single crystal [8]. The difference of spin polarization between single and poly crystals can be caused by the changing the mean free path l of the sample [9].

Finally, we show the temperature dependence of differential conductance between 2.0 K and 11 K at a Nb/ $SrRuO_3$ interface in figure 5. When the temperature is higher than the Nb superconducting transition temperature $T_C = 9.5$ K, the spectra show no anomaly. In contrast, as the temperature is decreased, a dip at around the zero bias voltage is grown rapidly because of the increase of the superconducting energy gap at Fermi surface. It is seen that the anomaly is well reproduced by the fitting curve calculated by the m-BTK function, representing fine control of the temperature. It is worthy to note that pumping a 1K pot to lower the temperature does not affect the differential conductance measurements, indicating that the spectrometer is mounted in ^{3}He cryostat or dilution refrigerator to perform the measurements at very low temperature region.

4. Conclusion

We have developed a "needle-anvil" type point-contact spectrometer to measure the spin polarization of ferromagnet down to $T \sim 2$ K. Two types of piezo devices, attocube piezo-based positioner and stacked type device, are used to control the contact size precisely between a sample ferromagnet and a superconductor. Owing to this method, the contact size between the tip and sample is varied from sub micro-meters to atomic-size contacts continuously. By suppressing thermal flow into the sample space and mechanical vibration, we can keep the contact over hours, enabling the precision measurements. To examine the performance of the spectrometer, we study the spin polarization of polycrystalline $SrRuO_3$ with the point-contact Andreev reflection measurements. The polarization is estimated to be ~0.59 at the clean limit of the interface, which is consistent with previous study.

Acknowledgments

This work was supported by a Grant-in-Aid for Scientific Research (Grant Nos. 25220605, 25287076, and 26600102) from the Japan Society for the Promotion of Science.

References

[1] Naidyuk G Y, and Yanson K I, Point-contact spectroscopy, 2004 Springer Series solid-state Sciences Vol. 145.
[2] Soulen J R Jr, Byers M J, Osofsky S M, Nadgorny B, Ambrose T, Cheng F S, Broussard R P, Tanaka T C, Nowak J, Moodera S J, Barry A, and Coey M D J 1998 *Sience* **282** 85
[3] Blonder E G, Tinkham M, and Klapwijk M T 1982 *Phys. Rev.* B **25** 4515
[4] Mazin I I, Golubov A A and Nadogorny B 2001 *J. Appl. Phys.* **89** 7576
[5] Kawasaki I, Yokoyama M, Nakano S, Fujimura K, Netsu N, Kawanaka H and Tenya K 2014 *J. Phys. Soc. Jpn.* **83** 0647712
[6] Strijikers J G, Ji Y, Yang Y F, Chine L C, and Byers M J 2001 *Phys. Rev.* B **63** 104510
[7] Shiga M, Nishimura N, Inagaki Y, Kawae T, Kambara H, and Tenya K, to be published in *J.Phys: Conf. Series.*
[8] Raychaudhuri P, MackenZie P A, Reiner W J and Beasley R M 2003 *Phys. Rev.* B 67 020411(R)
[9] Mazin I I 1999 *Phys. Rev. Lett.* **83** 1427

IOP Conf. Series: Journal of Physics: Conf. Series **897** (2017) 012002 doi:10.1088/1742-6596/897/1/012002

Low-temperature creep of austenitic stainless steels

R. P. Reed[1] and R.P. Walsh[2]
[1] Cryogenic Materials, Inc., Boulder, Colorado, U.S.
[2] National High Magnetic Field Laboratory, Tallahasse, Florida, U.S.

E-mail: rpreed@comcast.net

Abstract. Plastic deformation under constant load (creep) in austenitic stainless steels has been measured at temperatures ranging from 4 K to room temperature. Low-temperature creep data taken from past and unreported austenitic stainless steel studies are analyzed and reviewed. Creep at cryogenic temperatures of common austenitic steels, such as AISI 304, 310 316, and nitrogen-strengthened steels, such as 304HN and 3116LN, are included. Analyses suggests that logarithmic creep (creep strain dependent on the log of test time) best describe austenitic stainless steel behavior in the secondary creep stage and that the slope of creep strain versus log time is dependent on the applied stress/yield strength ratio. The role of cold work, strain-induced martensitic transformations, and stacking fault energy on low-temperature creep behavior is discussed. The engineering significance of creep on cryogenic structures is discussed in terms of the total creep strain under constant load over their operational lifetime at allowable stress levels.

1. Introduction

Even at stresses below the yield strength, limited creep has been demonstrated at very low temperatures in metals and alloys, including austenitic stainless steels. Austenitic steels are being used extensively in cryogenic equipment, cryogenic liquid storage and transportation, and for encasement and structural support of superconducting magnets. This paper includes three aspects related to the creep of austenitic stainless steels at low temperatures: (1) unreported creep measurements at room temperature, 76 K, and 4 K for three austenitic stainless steels (AISI 304HN, 310, and 316), (2) review of all reported cryogenic creep data for austenitic stainless steels, and (3) estimates of expected creep strains for austenitic steels under typical design criteria of structural alloy codes at low temperatures. Discussion is included in the review section related to creep deformation mechanisms and the role of cold work, strain-induced martensitic transformations and stacking fault energy in affecting creep in austenitic stainless steels at low temperatures.

2. Creep measurements

2.1. Materials:

Three annealed austenitic stainless steels were included in this test series: AISI 304HN, AISI 310, and AISI 316. Their chemical composition, grain size, hardness, tensile yield strength (σ_y, 0.2% offset), and elastic moduli at 295, 76, and 4 K are listed in Tables 1 and 2. With the exception of the elastic modulus, all of the parameters listed in Tables 1 and 2 were measured for the same production heats of the alloys on which creep tests were conducted in this test series. At 4 K the elastic moduli for alloys 304HN and 316 are lower than at 76 K owing to the presence of the Neel temperature (para-to-antiferromagnetic transition) in these alloys below 76 K. Alloy 304HN has very high nitrogen content and is stable with respect to low-temperature, strain-induced martensitic transformations. Alloy 310 contains a high nickel content which leads to increased stability of the face-centered cubic austenite phase, thus to suppression of the strain-induced martensitic transformations. Alloy 316 has a metastable austenitic structure and partially transforms to martensite during plastic deformation at low temperatures. Other low-temperature mechanical property data from the same production heats of these alloys have been previously reported [1 -6].

Round tensile specimens were cut and machined from annealed 20 mm diameter alloy 316 and 310 bar stock and from annealed 12.5 mm thick 304HN plate. The reduced sections of the specimens were 6.3 mm in diameter.

Table 1. Chemical composition, grain size, and hardness of austenitic steels

AISI alloy	chemical composition*, wt. %								grain size	hardness
	Cr	Ni	Mn	Mo	C	N	Cu	Si	(μm)	(R_B)
304HN	20.0	8.4	-	0.6	0.065	0.264	0.5	--	45	--
310	24.8	20.8	1.7	0.1	0.093	0.031	0.1	0.7	161	72
316	17.2	13.5	1.9	2.3	0.057	0.03	--	0.58	65	79

* S and P content of all alloys less than, or equal to, 0.02 wt.%

Table 2. Tensile yield strength of three annealed austenitic steels used in this study

AISI alloy	tensile yield strength (0.2% offset), MPa		
	295K	76K	4K
304HN	380	975	1240
310	225	545	760
316	240	610	660
	elastic modulus, GPa		
304HN	200	214	210
310	191	205	207
316	195	209	208

2.2. Test procedures

Strain-gage (Fe-73Ni-20Cr, metal-film, resistance) extensometers were used to sense plastic deformation (strain) for each creep test. Their strain sensitivity was 5 x 10^{-7} at all test temperatures. Over a period of one month, strain variability from instrumental and temperature drift ranged from about +/- 3 x 10^{-7} at room temperature to about +/- 4 x 10^{-6} in liquid helium. Additional discussion of low-temperature measurement techniques and inconsistencies is contained in studies of low-temperature indium and copper creep [7-8].

Servo-hydraulic testing equipment was used to apply and maintain constant load during each creep test. The load sensitivity of this commercial equipment was +/- 0.1 % and load variability was +/- 2 %. An inner titanium pull-rod and outer glass/epoxy compression cylinder were used in the load train. The super-insulated dewar capacity was 30 liters (900 mm depth) and the boil-off rate of liquid helium during testing was 0.5 liters per hour.

2.3. Test results

The strain identified in low-temperature creep of austenitic stainless steels included (1) the initial elastic deformation that occurred during loading to the desired stress level; (2) an immediate initial plastic deformation (arbitrarily selected as the first 6 seconds of each test of the application of the dead-weight load) that is termed the primary creep stage; (3) subsequent long-term plastic deformation (creep) that is termed the secondary creep stage. (4) the final stage, termed tertiary, identifies either creep strain-rate acceleration (leading to sample failure) or gradual reduction of the creep rate, ending in total suppression (exhaustion) of creep strain. Exhaustion has been observed during creep tests at 4 K and these cases are identified in discussion and figures. Creep acceleration, leading to failure, was not been observed in this test series.

The secondary creep stage consisted of a logarithmic dependence of creep strain on elapsed time (t) at all test temperatures (4 K – liquid helium, 76 K – liquid nitrogen, and 295 K – room temperature). All creep tests were conducted at stresses (σ) in the vicinity (σ/σ_y = 0.6 to 1.35) of the alloy tensile yield strength. In all tests at applied stress/yield strength ratios above 0.73 some plastic deformation was recorded.

Creep data from these measurements are plotted in Figure 1 as creep strain/Δ ln time versus normalized applied stress where creep strain represents the total secondary-stage strain during the time period from 0.1 to 4 x 10^4 minutes. Here, Δ ln t = ln [4 x 10^4 min x 60 (s/min)] – ln 6 (s), and the normalized applied stress = applied stress/yield strength (provided in Table 2). In Figure 1 and in subsequent figures, the normalized applied stress (σ_y/σ) is plotted, instead of the applied stress. This ensures better comparison of data between different test temperatures and different alloys. The data at 4 K are magnified in Figure 1 (right side) for better clarity. Notice that at higher normalized stress levels the secondary-stage creep strains for alloys 310 and 316 are suppressed to practically zero. The data at room temperature and 76K are equivalent (within data scatter), but considerably larger than the creep rates were measured at 4 K. The secondary-stage creep rate at 76 K of alloy 316 is very slightly lower than alloy 310 and 304HN rates at 76K - which are equivalent.

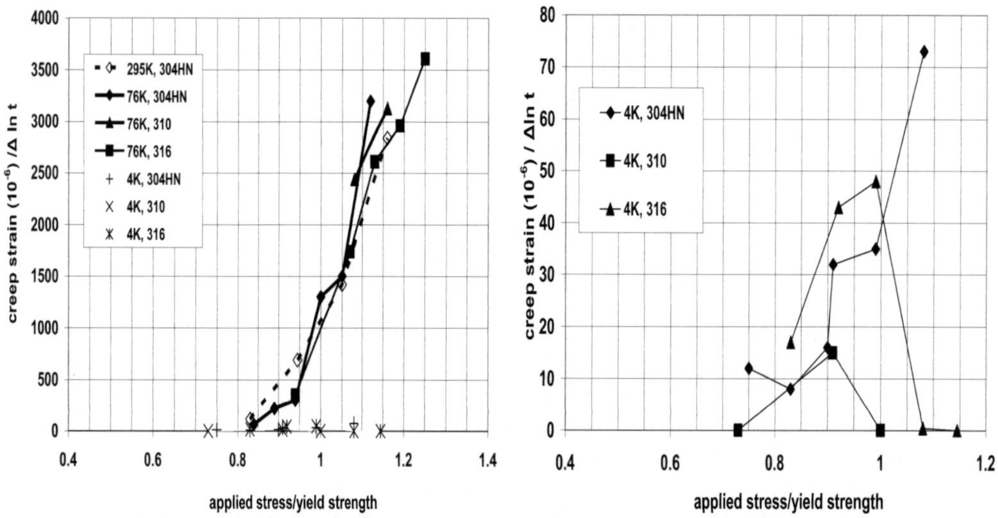

Figure 1. Secondary-stage creep strain/Δln t from 0.1 to 4 x 10^4 minutes for annealed austenitic stainless steel alloys 304HN, 310, and 316 plotted as a function of normalized applied stress. Left side: for 295, 76 and 4 K; right side: for 4 K

Figures 2-4 present data for the total creep strain (primary plus secondary stages), and secondary-stage creep strain for all alloys at 4, 76, and 295 K for an intermediate test duration of 1200 minutes. This time was selected to permit comparison of all test data at an equivalent time. In Figure 2 the total primary stage (including elastic strain) plus secondary-stage creep strains are plotted versus normalized applied stress. All alloys have identical dependences at 295, 76, and 4K test temperatures of total strain on normalized applied stress. In Figure 3 the secondary-stage creep strains within the time interval of 0.1 to 1200 minutes for all alloys at all temperatures are plotted. Notice the very low strains at 4 K, the lack of distinctions between alloys of the dependence of creep strain on normalized applied stress, and the similarity between room temperature and 76 K data for alloy 304HN. The relatively low secondary-stage strains at 4 K are amplified for view in Figure 4. The alloy 316 data indicate suppression (exhaustion) of creep processes at normalized applied stresses above 1.2.

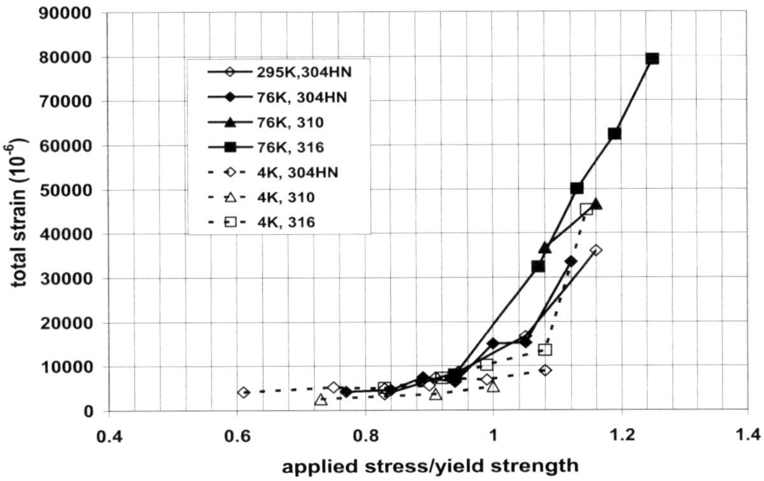

Figure 2. Total specimen strain (primary + secondary) for test durations of 1200 minutes at 295, 76, and 4 K for annealed austenitic stainless steel alloys 304HN, 310, and 316.

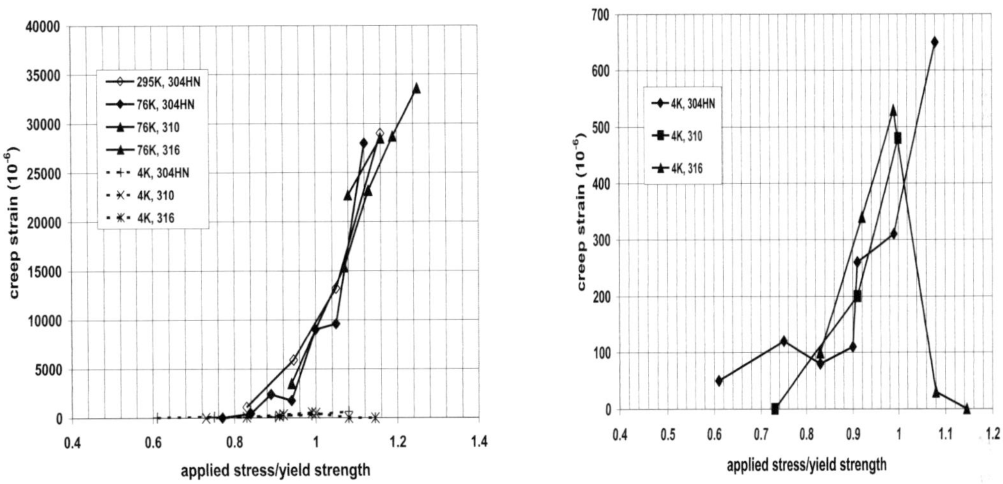

Figure 3. Creep strain (secondary stage) from to 1200 minutes for annealed austenitic stainless steel alloys 304HN, 310, and 316. Left side: at 295, 76, and 4 K; right side: at 4 K, magnified strain

Creep strain rates can be estimated for each alloy at each test temperature. In Table 3 these rates are presented for applied stress/yield strength ratios as close to 1.0 as were tested after test durations of 10^4 minutes.

Table 3. Creep strain rates for annealed austenitic steels at low temperatures

Alloy	temperature, K	applied stress/yield strength	creep rate (10^{-10} s^{-1})
304HN	295	1.0	67
	76	1.0	58
	4	1.0	1.7
310	76	1.08	83
	4	1.00	1.7
316	76	1.15	48
	4	1.0	1.

2.4. Discussion

At low temperatures the total creep strain (ε_c) at applied stresses in the vicinity of the yield strength consists of three stages: primary – ε_p (linear elastic, then parabolic plastic strain dependence on time), secondary plastic creep strain (logarithmic dependence on time) – ε_s, and tertiary – ε_t (either an increase of strain dependence on time – leading to failure or a steady decrease of strain dependence on time - leading to total suppression, exhaustion, of the creep process). The total creep strain is:

$$\varepsilon_c = \varepsilon_p + \varepsilon_s + \varepsilon_t \tag{3}$$

The primary and secondary stages of creep at room temperature, 76, and 4 K for three annealed steels (304HN, 310, and 316) have been characterized in this study. Data were presented in Figure 1 depicting logarithmic creep dependences ($\varepsilon_s/\Delta \ln t$) versus normalized applied stress (applied stress/yield strength) within the secondary creep stage for 295, 76 and 4 K where t is the test time increment within the secondary creep stage. Total amounts of creep strain for the combined primary and secondary creep stages at each test temperature are included in Figure 2 where the primary stage duration was 0.1 minutes (6 s) and the secondary stage time: 7.2 x 10^4 s – 6 s. In Figure 3 only the secondary-stage creep strain is shown over the same time range as in Figure 3. The following general conclusions can be drawn from the creep data presented in Figures 1-3.

2.4.1. Primary creep

At room temperature, 76 K, and 4 K the primary stage of creep in austenitic steels follows closely the shape of the stress –strain curves. Below the elastic limit (at normalized stress levels about 0.75) there is linear elastic strain; above the elastic limit approximate parabolic plastic deformation dominates.

2.4.2. Secondary creep

Data presented in Figures 1-3 indicate the following:

* There is the same dependency of secondary-stage creep strain on normalized applied stress for all alloys at all test temperatures. This dependency has not been previously reported and suggests a single creep mechanism for all test conditions.

* The secondary creep strains at 4 K for the three alloys are significantly less compared to 76 and 295 K. Creep exhaustion during the secondary creep stage was found in two of the three alloys.

3. Review of creep data

An extensive review of the low-temperature creep of metals and alloys has been presented by Tien, et al [9]. This review primarily focused on creep measurements and mechanisms of elements at low temperatures. At the time of the review there had not been a comprehensive study of creep in austenitic steels, but a portion of the Voyer, Mugnier, Menard, and Weil [10-13] studies on alloys 304L, 304LN, and a 310-similar alloy were briefly included. They conducted low-temperature creep measurements (using an Andrade-type, external optical lever-arm for displacement sensing) at 300, 77, and 20 K and studied the role of the austenite-martensite transformation on the creep deformation processes of 18Cr-10Ni (304Land 304LN) and 25Cr-20Ni (310) steels. Demchuk, et al [14] reported on the creep of a stable 25Cr-20Ni (310) austenitic alloy at 77 K, also using an Andrade-type, displacement-sensing assembly. Roth, et al [15] used capacitance gages to measure the creep of a 304LN alloy at 4 K. Ogata, et al, [16-17], using strain-gage extensometers and a servo-hydraulic load control equipment, measured the creep of alloys 304L, 310S, and 316LN at room temperature, 77 K, and 4 K. More recently, Usami and Mori [18] have measured creep at 293 and 77K of cold-worked alloy 304 and annealed alloys 316L and 316LN austenitic steels using strain gages and load-control equipment.

Earlier low-temperature studies (except Ogata, et al [16-17] and Osami and Mori [18]) focused on measurement of the plastic deformation following application of dead-weight loads. In all studies the measured creep strain has been observed to be linearly dependent on the log of the test

IOP Conf. Series: Journal of Physics: Conf. Series **897** (2017) 012002 doi:10.1088/1742-6596/897/1/012002

time and referred as "logarithmic creep". Creep-strain exhaustion was observed at 4 K. The other possibility for the tertiary creep stage, creep strain acceleration leading to sample failure, has been little studied owing to the requirement of long durations at low temperatures.

Low-temperature creep tests are difficult to conduct since mechanical, thermal, and electrical stabilities are required within a cryostat surrounding the specimen for extensive periods of time. Cryogen boil-off rates must be minimized, refill procedures carefully controlled, and liquid levels accurately monitored. Strain sensors typically have instrumental drifts that must be minimized. During operation, thermal contraction of load-train components, ice formation along seals for the central load trains, as well as small-scale external electrical, mechanical, and vibration irregularities lead to small increases or decreases in applied load and/or sample micro-strains. These inconsistencies have resulted in a fairly large spread of low-temperature creep data.

A summary of the creep data for austenitic stainless steels at room and low temperatures is presented in Figures 4-7. All of the alloys in these figures were tested in the annealed condition except for alloy 304 of the Usami and Mori study [18]. This alloy was extensively cold worked (yield strength at 293K = 630MPa, compared to the annealed yield strength of this alloy of 290 MPa). The duration of the creep measurements of Weil, et al [10-12] were considerably shorter than the test times of other measurements, therefore a shorter time duration (0 1 to 1200 minutes) of the secondary creep stage was used for comparison of data in Figures 4-5. In Figure 4 it is apparent that the slopes of the Weil, et al data are lower than other data. Also, the Weil creep data were obtained on some samples under much larger applied stress levels at all test temperatures. The data obtained at 20 K (Figure 5) has a similar dependence of creep strain on normalized applied stress as the higher temperature data shown in Figure 4. At both 77 K and 20 K, the Weil creep strain data for alloys 304L and 304LN at larger normalized applied stress levels decrease, but their alloy 310 creep strain continues to increase. This reduction of creep rates at higher stress levels is attributed to the presence of strain-induced martensite in the less-stable 304-base alloys. This is discussed additionally in section *3.1.* There is good agreement between the 4 K test data as shown in Figures 5b and 7. Note the suppression found at higher stress levels found in some alloys after 1200 minutes and most alloys after 4 x 10^4 minutes.

There are insufficient data in these studies on the primary creep stage and no information related to the tertiary creep stage which demands excessively long test times for completion. Information related to these two creep stages are not included in this review.

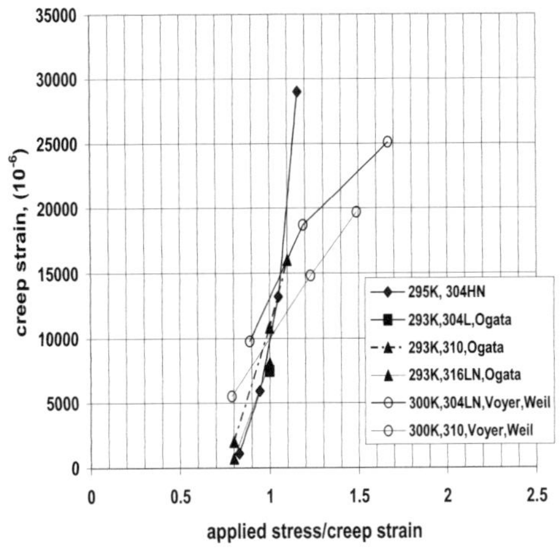

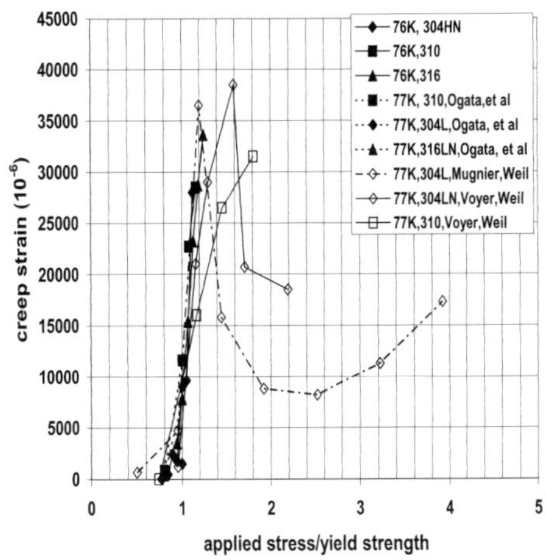

Figure 4. Strain during secondary creep stage from 0.1 to 1200 minutes for annealed austenitic steel alloys from this study, Ogata, et al [16,17], Voyer and Weil [10-11] and Mugnier and Weil [12]. Left side: room temperature; Right side: at liquid nitrogen temperature.

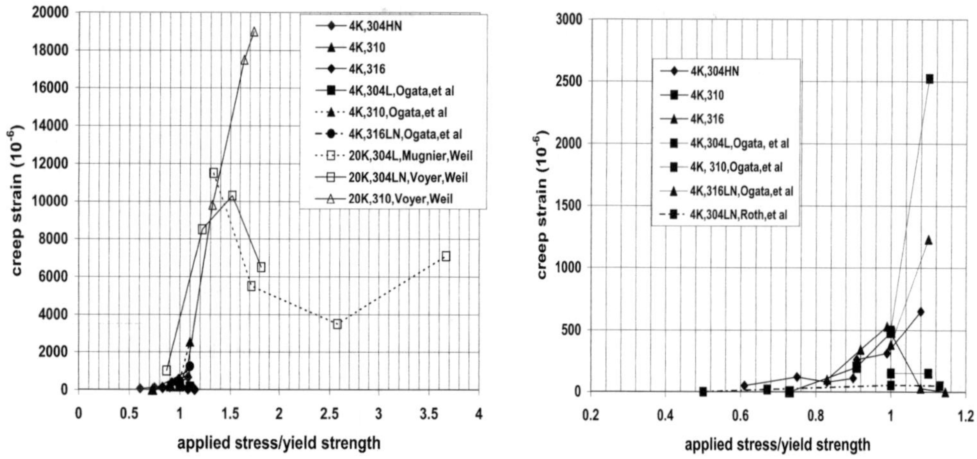

(a) **(b)**

Figure 5. (a) Strain during secondary creep stage from 0.1 to 1200 minutes for annealedaustenitic steels at 20 and 4 K from this study, Ogata, et al [16-17], Voyer and Weil [1—11], and Mugnier and Weil [12]. **(b)**. at 4 K from this study, Ogata, et al [16-17], and Roth, et al [15].

The creep strain/Δln$t$ for the secondary stage is plotted versus the normalized applied stress over the longer test duration of 0.1 to 4x10^4 minutes at 293-295 and 76-77 K (Figure 6), and at 4 K (Figure 7) for available test data (this paper and references 10-18). Agreement between the data from this paper, Ogata, et al and Roth, et al is very good. However, the creep-strain data from Osami and Mori are slightly less than other data at room temperature and slightly more than other data at 77 K. The data from this paper, Ogata, et al, and Roth, et al, when plotted as secondary-stage creep strain versus normalized applied stress, are independent of temperature. However, the creep-strain data of Osamu and Mori at room temperature are higher than their data at 77 K. There is little distinction between alloy types in all test data. All data indicate the same dependency of creep strain / Δln$t$ on normalized applied stress. This suggests that the creep mechanism is independent of test temperature and of each alloy within the secondary creep stage. The magnitude of the creep strain/Δln$t$ values at both 293-295 K and 76-77 K are significantly larger than those at 4 K. At 4 K all alloys show some evidence of exhaustion of the creep strain at larger values of the normalized applied stress.

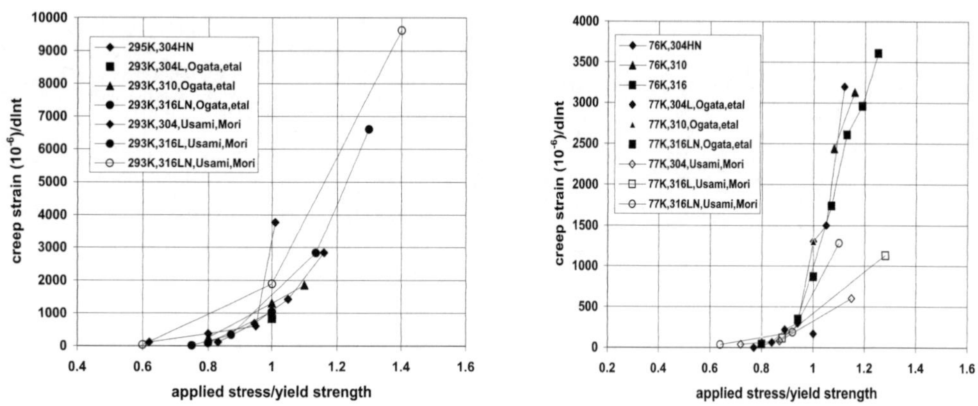

Figure 6. Secondary creep strain/Δln t from 0.1 to 4 x 10^4 minutes from this_study and others [16 -18], versus normalized applied stress. Left: at room temperature. Right: at 76-77 K

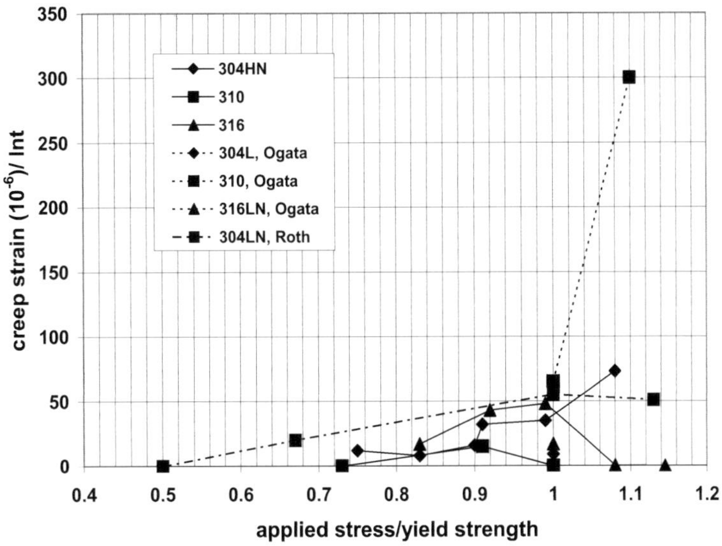

Figure 7. Secondary creep strain/Δln t at 4 K from 0.1 to 4 x 10^4 minutes for austenitic steel from this study, Ogata, et al [16-17], and Roth, et al [15] plotted versus normalized applied stress

In the following sections of this review, the possible effects from martensitic transformations, variations of the stacking fault energy and prior cold work, and creep mechanisms are discussed.

3.1. Martensite effects

Martensitic phase transformations in austenitic stainless steels have been extensively studied and have been reviewed by Reed [19-20]. The crystal structure of austenitic stainless steels in the annealed state at room temperature is face-centered cubic (fcc) and during plastic deformation at cryogenic temperatures some alloys partially transform to a hexagonal-close packed (hcp) phase and at larger deformations transform to a body-centered cubic (bcc) phase. The transformations are diffusionless, involve the shear of atoms at localized sites, and are labeled as martensitic transformations [19-20]. In common stainless steels the austenitic (fcc) phase is stable with respect to cooling to 4 K, but many of these alloys are susceptible to partial transformation to the martensitic products during normal tensile deformation at room temperature and lower temperatures. The fcc-to-hcp transformation is associated with deformation faulting during early stages of low-temperature deformation and leads to an "easy glide" region in stress-strain curves of low stacking fault energy austenitic steels [20]. The temperature at which the more common fcc-to-bcc martensitic transformation first initiates on cooling is labeled as the martensite start (M_s) temperature and at which martensite first initiates during plastic deformation is labeled as the martensite deformation (M_d) temperature.

Base alloys that contain modest amounts of austenite-stabilizing elements such as Ni, Mn, Mo, N, and C are thermodynamically metastable at low temperatures while additional amounts of these elements serve to increase austenite stability and, thus, to decrease the tendency of strain-induced martensitic transformation. Of the set of alloys that have been previously measured for low-temperature creep (discussed above) alloys 304 and 316 with lower Ni and N contents, tend to transform, partially, to martensitic products during plastic deformation at low temperatures and alloy 310, with higher Ni content, is stable with respect to strain-induced martensitic transformation.

The calculated averages and ranges of the M_d and M_s temperatures are listed for selected austenitic stainless steels of this study in Table 4 from past work [21-28]. Stacking fault energy predictions are also included in Table 4; lower stacking fault energies lead to additional deformation faulting and to more extensive fcc-to-hcp martensitic transformation during low-temperature deformation. Notice that the M_d temperatures of 316 and 304 type alloys are above the creep test temperatures of this study; therefore one would expect some martensitic transformation during creep for these two alloys.

Table 4. Parameters related to the fcc-to-bcc martensitic transformation and stacking fault energy in austenitic stainless steels [21-28]

Alloy	temperatures, K martensite start during: cooling(M_s)	deformation (M_d)	stacking fault energy mJ/(m^2 x wt.%)
304HN	- 440	+380 (261-498)	43 (34-54)
316	- 354	+220 (183-262)	64 (61-70)
310	-1660	- 390 (-361-418)	84 (73-94)

The strain-induced transformation from fcc-to-bcc structures is associated with increased work hardening during deformation at low temperatures. The creep data of Voyer, Mugnier, Menard, and Weil [10-13] that are included in Figures 4-5 indicate the role of strain-induced bcc martensitic formation in the suppression of creep strain at higher stress levels during tests at 77 and 20 K. The decrease of creep strain at higher stress levels is shown in these figures for their alloys 304L and 304LN, but not for the stable 310 alloy. There is no reduction, but a slightly lower slope, of the creep strain at higher normalized applied stresses in the room temperature data; this likely corresponds to the formation of much less strain-induced bcc martensite at room temperature. These authors include electron microscope pictures of the deformed austenite for their alloy 304L that clearly shows the presence of bcc martensite in specimens that were exposed to creep at higher stress levels at 77 and 20 K. The creep data of other investigations that are included in Figures 4-7 were not obtained at comparably high normalized applied stress levels.

3.2. Cold work effects

Only one of the alloys included in this review had been cold-worked prior to testing – the 304L included in the Usami, Mori study [18]. The alloy was heavily cold worked, presumably at room temperature, and, therefore, contained a sizable amount of strain-induced martensite. For this alloy the M_d temperature is slightly higher than the value of 380 K listed for alloy 304HN in Table 4, owing to the reduced amounts of N and C. Unfortunately, the creep strain data reported for this alloy are inconsistent as to the effect of added cold work and bcc martensite in this alloy: at 293 K the data point at the normalized applied stress level of 1.02 is higher than all other data, but the other 5 data points at 293 and 77 K in Figures 10 and 11 are all lower than equivalent data (which would be predicted from the presence of substantial bcc martensite in the samples).

It is likely that the presence of cold work does not alter the dependence of creep strain on normalized applied stress. But, since cold work increases the yield strength, cold work will result in lower *normalized* applied stresses under the same operating stress conditions and, thus, result in lower creep strains under these conditions.

3.3. Stacking fault energy effects

It is commonly thought lower stacking fault energies restrict creep strain. Lower fault energies are associated with wider separation of dissociated partial dislocations, commonly present in close-packed crystal structures, such as austenitic steels. More widely separated partial dislocations of both mobile and forest dislocations act to restrict jog formation of these dislocations, a lower-energy process which enables easier passage of mobile dislocations through the forest dislocations during creep. The discussion below suggests that this in not necessarily the creep process for austenitic steels at low temperatures.

The low-temperature stress-strain curves of low stacking fault energy, austenitic stainless steels have a distinct "easy glide" range that immediately follows the initial plastic deformation associated with the yield strength. Plastic deformation within this range has been associated with extensive deformation faulting and fcc to hcp martensitic transformation along {111} slip planes. One would expect to see evidence of this extensive faulting during creep tests of these low stacking fault energy alloys.

Stacking fault energies for a range of austenitic stainless steels have been measured using electron microscopy (dislocation node sizes) and x-ray diffraction (peak shifts) and reviewed by Reed [20] and Novak [28]. Average calculated values for the alloys included in this measurement series are presented in Table 4. The values of Table 4 are for room temperature; the stacking fault energy is expected to decrease at lower temperatures for austenitic steels. The "easy glide" region of the stress-strain curves of 304 base alloys, especially significant at 77 K, has been associated with the lower stacking fault energy of this alloy series at this temperature [20} and has been attributed to extensive deformation faulting and hcp martensite transformation along the primary {111} slip plane. Lower stacking energy alloys of this test series and those that are included in this review should also have deformation faulting and hcp martensite contributions to their creep deformation for 76-77 K tests.

From Figures 4-7, there appears to be no distinction of the creep strain versus normalized applied stress between the low stacking fault energy alloy 304 and the high stacking fault energy alloy 310. Yet, careful comparisons between the three alloy types from each investigation are listed in Table 5. In Table 5 ranges of the stacking fault energy for each alloy class, extending between 295 – 77 K and data for secondary-stage creep strain $(10^{-6})/\Delta \ln t$ at a normalized applied stress ratio of 1 for each type of alloy from each study are listed. Notice in the table that in only two of the seven data sets, both at room temperature, do the logarthimic creep strain trends decrease with increasing stacking fault energies. This lends considerable support to the premise of modest dependence of creep strain on stacking fault energy.

Table 5. Comparison of stacking fault energy ranges with logarithmic creep rates for austenitic steels

	304	**316**	**310**
Stacking fault energy range (mJ/m^2)	20/50	60/70	75/95
Secondary creep strain $(10^{-6})/\Delta \ln t$ at $\sigma/\sigma_y = 1$			
This study, 76 K	1000	1000	1800
Ogata, et al, 293 K	830	1040	1300
77 K	170	870	1300
Usami, Mori, 293 K	3700	1500, 1600	----
77 K	300	600, 700	----
Voyer, Weil, 300 K	1300	-----	1000
77 K	700,1000	-----	1000

3.4. Creep Mechanism(s)

Previously, data for low-temperature creep has usually been portrayed by plotting the creep strain/log time versus applied stress (see, eg, Tien and Yen[1]) and a linear dependence between those variables has been shown for each test temperature. In this paper, secondary-stage creep strain/log time is plotted versus the normalized applied stress (applied stress/yield strength). Within experimental uncertainties, the same trend for data at all test temperatures (295 - 4 K) for all alloys is demonstrated. This strongly suggests that the same creep deformation mechanism is active at all test temperatures.

However, this study encompasses a very diverse range of deformation processes that include dislocation generation and motion in the more stable austenitic stainless steels such as alloy 310 and deformation faulting, hcp-to-fcc martensitic transformation, deformation twinning, and small amounts of bcc martensite in the metastable alloys 304 and 316. Deviations from this trend were only observed (at high normalized applied stress rations) when strain-induced bcc martensite (corresponding to that formed in strong work-hardening stage of their stress-strain curves [10.11]) was present. The presence of this type of martensite has been found to decrease, considerably, the creep strain dependence on applied stress.

Reduction and/or exhaustion of the creep process was observed at 4 K at normalized applied stress levels above about 1.

4. Design effects

Most design criteria of cryogenic structures stipulate that the primary operating stresses not exceed 50% of the yield strength at either room temperature or the operational temperature. At normalized applied stress levels at low temperatures for these cases, there would be no detectable secondary-stage creep strain. However, some design criteria permit primary operating stresses up to 75% of the yield strength. In this case at 295 and 77 K, there would be very small amounts of secondary-stage creep strain over a 20year operation lifetime and more sizable initial, primary-stage strain. For an operating stress equal to the yield strength at the operating temperature a more sizable total strain would be incurred during a 20 year lifetime. At 4 K only the primary stage strain would be observed for any applied stress; secondary-stage creep strain is insignificant (but estimated) at this temperature. Quantitative estimates of the primary and secondary creep stage strains for a 20-year operating lifetime are listed in Table 6. Consider that the elapsed time for the primary stage is only about 6 seconds while the elapsed time for the secondary stage is 20 years.

Table 6. Plastic strain during creep over an operational lifetime of 20 years

Stress level	Temperature, K	Primary stage	Secondary stage	Total strain
$<1/2\ \sigma_y$	295 - 4K	0	0	0
$<3/4\ \sigma_y$	295 K	0.005	0.003	0.008
	76-77 K	0.005	0.002	0.007
	4 K	0.001	< 0.0003	< 0.0013
σ_y	76–77 K	0.015	0.020	0.035
	4 K	0.002	< 0.0006	< 0.0026

5. Conclusions

The creep characteristics of three austenitic stainless steels were measured at low temperatures and the low-temperature creep of austenitic steels is reviewed with special attention to the effects of cold work, martensitic transformations, and stacking fault energy. The results are summarized here.

All secondary-stage creep data show similar dependence of the creep strain/$\Delta \ln t$ on normalized applied stress for the temperature range 4 – 300 K for all austenitic stainless steels. This dependency has not been previously reported and suggests a single creep mechanism for austenitic stainless steels under the test conditions of these studies.

Cold work raises the yield strength, thus lowers the creep strain at the same applied stress level.

The formation of strain-induced bcc martensite at normalized applied stress levels above 1.1 significantly suppresses subsequent secondary-stage creep strain.

There is a modest effect of increased stacking fault energy on increased creep strain in austenitic steel alloys at 77 K

Creep exhaustion occurs at 4 K in secondary-stage creep at normalized applied stress levels above about 1.0.

IOP Conf. Series: Journal of Physics: Conf. Series **897** (2017) 012002 doi:10.1088/1742-6596/897/1/012002

Design criteria with primary operating stress levels limited to less than ½ of the yield strength will preclude detectable creep strain during 20-30 year operation.

Acknowledgements

The creep tests were conducted by the authors at the cryogenic mechanical property laboratory at the National Institute of Standards and Technology (NIST) and Materials Research and Engineering, Inc., Boulder, Colorado prior to the authors' departures from NIST.

References

[1] Reed R, Simon N, Walsh R 1990 *Advances Cryogenic Engineering-Materials* **36B** pp 1175-1183
[2] Reed R, Simon N 1992 *Advances Cryogenic Engineering-Materials* **38A** pp 117-126
[3] Reed R, Simon N 1989 *HNS 88 – High Nitrogen Steels, Institute of Metals*, London pp 111-188
[4] Reed R, Walsh R 1988 *Advances Cryogenic Engineering-Materials* **34** pp 199-208
[5] Reed R, Simon N, 1990 *Advances Cryogenic Engineering-Materials* **36B** pp 1077-1086
[6] Tobler R. Reed R, Burkhalter D 1980 *Advances Cryogenic Engineering-Materials* **26** pp 107-119
[7] Reed R, Walsh R 1992 *Advances Cryogenic Engineering-Materials* **38A** pp 117-126
[8] Reed R, Simon N, Walsh R 1991 *Materials Science Engineering,* **A147** pp 23-32
[9] Tien J, Yen C-T 1984 *Advances Cryogenic Engineering – Materials* **30** pp 319-338
[10] Voyer R, Weil L 1965 *Advances Cryogenic Engineering* **10** pp 110-116
[11] Voyer R, Weil L 1966 *Advances Cryogenic Engineering* **11** pp 447-452
[12] Mugnier D, Weil L 1963 *Advances Cryogenic Engineering* **9** pp 98-103
[13]. Menard J, Weil L 1961 *Advances Cryogenic Engineering* **6** pp 587-589
[14] Demchuk I, Krakhmalev V, Lebedev D 1979 *Metal Sci. Heat Treatment* **21** pp 547-553
[15] Roth L, Manhardt A, Dalder E, Kershaw R 1986 *Advances Cryogenic Engineering-Materials* **32** pp 369-376
[16] Ogata T, Ishikawa K 1986 *Cryogenics* **26** pp 365-369
[17] Ogata T, Umezawa O, Ishikawa K 1990 *Advances Cryogenic Engineering-Materials* **36B** pp 1233-1239
[18] Usami S, Mori T 2000 *Cryogenics* **40** pp 117-126
[19] Reed R 1983 Chapter 9, *Materials at Low Temperatures*, R. Reed, A. Clark editors, American Society for Metals, Metals Park, OH pp 295-341
[20] Reed R 1988 "Austenitic Stainless Steels with Emphasis on Low-Temperature Strength", *Alloying*, Chapter 9, J. Walter, M. Jackson, C. Sims editors, ASM International, Metals Park, Ohio, pp 225-256
[21] Hull F 1973 *Welding Journal* **52** pp 193-203
[22] Eichelman G, Hull F 1953 *Transactions American Society Metals* **45** pp 77-104
[23] Self J 1986 "Effects of Composition uoon the Martensite Transformation Temperature of Austenite Steel Welds", Center for Welding Research, Colorado school of Mines, Golden, CO
[24] Williams I, Williams R, Capellano R 1976 *Proc, Sixth International Cryogenic Engineering Conference,* IPC Science and Technology Press, Guildford, Surrey, England pp 337-341
[25] Schramm R, Reed R 1975 *Metallurgical Transactions* **6A** pp 1345-1351
[26] Rhodes C, Thompson A 1977 *Metallurgical Transactions* **8A** pp 1901-1906
[27] Brofman B, Ansell G 1978 *Metallurgical Transactions* **9A** pp 879-880
[28] Novak C 1977 "Structure and Constitution of Wrought Austenitic Stainless Steels" *Handbook of Stainless Steels*, Chapter 4, D. Peckner, I. Bernstein editors, McGraw-Hill Book Company, New York pp 4-1 to 4-78

IOP Conf. Series: Journal of Physics: Conf. Series **897** (2017) 012003 doi:10.1088/1742-6596/897/1/012003

Mechanical properties of $LaFe_{11.5}Si_{1.5}$/Cu negative thermal composite and its application as clamp materials for tensile test at cryogenic temperatures

Shan Y,[1,2] Huang R,[1,2] Zhao Y,[1,2] Huang C[1] and Li L[1,2]

[1]Key Laboratory of Cryogenics, Technical Institute of Physics and Chemistry, Chinese Academy of Sciences, Beijing, P. R. China

[2]University of Chinese Academy of Sciences, Beijing 100049, P. R. China

E-mail: huangrongjin@mail.ipc.ac.cn

Abstract. La (Fe, Si)$_{13}$ compounds have been widely studied for their excellent negative thermal expansion (NTE) properties. However, their poor mechanical properties limit their practical applications. In this work, $LaFe_{11.5}Si_{1.5}$/Cu material was fabricated. It was found that the NTE behavior occurs obviously at cryogenic temperatures and the ratio of $\Delta L/L$ can reach to 0.12%. Mechanical tests indicated that the absolute value of compressive strength at 77K and 300K is 365MPa and 222MPa, respectively. The elastic modulus at 77K and 300K is -109GPa and -87GPa, respectively. In addition, the average hardness is 337Hv performed in the Vickers hardness tester. Loose between the samples and clamps in the tensile test due to the contraction of clamp at low temperatures remains a big issue. In order to solve this problem, some $LaFe_{11.5}Si_{1.5}$/Cu NTE sheets are added between the clamp and the tested samples. Results showed that the samples with NTE materials sheets embedded is held tighter by the clamp. The maximum force of the tensile test is 9.77N and 5.48N, respectively, which illustrates that the adding of NTE material does make sense.

1. Introduction

In common sense, the volume of most materials can be changed through temperature shift, obeying a principle of metal expanding on heating and contracting on cooling, which is so-called normal thermal expansion. However, this property easily leads to fatal error in cryogenic environment, especially in aerospace, precision instrument, mechanical device and cryogenic engineering. In late 19th Century, abnormal thermal expansion material was first found by Guillaume[1]. Invar, with the constitution of Ni, Fe and few C, its coefficient of thermal expansion (CTE) is very low and expand hardly with the change of temperature. In 1951, Hummel found that the crystalline aggregate of β-eucryptite, while in an environment over 1000℃, it contacts with temperature increasing[2]. This abnormal phenomenon is called negative thermal expansion (NTE).

Until now, some materials with excellent NTE properties have been successfully fabricated and tested, such as ZrW_2O_8[3, 4], PbTiO3[5], ScF_3[6, 7], antiperovskite manganese nitride[8, 9] and La(Fe, Si)$_{13}$-based compounds[10]. Cubic La(Fe, Si)$_{13}$-based compounds, as one of the typical magnetic refrigeration materials, have been deeply studied by researchers[11-13], which exhibit large, isotropic and nonhysteretic NTE properties as well. Researches also showed that, by changing the amount of Si, doping other elements into （La(Fe, Si)$_{13}$-based compounds, better NTE properties can be achieved[14].

Though NTE materials have been optimized, the mechanical properties of most NTE materials have not be considered thoroughly for fragile materials easily break in application[15]. Cu, with high ductility, its adding in La(Fe, Si)$_{13}$-based compounds is expected to exhibit good NTE and mechanical properties in practical application. Studies show that the doping of Cu can effectively enhance the mechanical properties of material and significantly affect the CTE of La(Fe, Si)$_{13}$-based compounds[16].

In the field of material science, mechanical test is necessary to evaluate property of certain material. However, in low temperature environment, the clamp used in the tensile test easily loose due to its contraction at cold temperatures, which makes the test results inaccurate. To solve this problem, NTE materials were considered to be added on the sample. Its good property—expanding at cold environment is expected to make the sample held tighter when the temperature descends.

In the previous study, Wang et al.[14] had tested the NTE properties of $LaFe_{13-x}Si_x$ (x=1.5, 1.8, 2.1 and 2.4) and found that $LaFe_{11.5}Si_{1.5}$ possess the biggest absolute value of average CTE ($-5.6\times10^{-6}K^{-1}$). In our work, 20 wt.% of Cu powder was added into $LaFe_{11.5}Si_{1.5}$. Its mechanical properties have been studied and this material is placed high hopes on to improve clamp's performance in cryogenic environment.

2. Experiment

2.1. *Synthesis of material*

Polycrystalline samples of $LaFe_{11.5}Si_{1.5}$ were prepared in an arc-melting furnace under a high-purity argon atmosphere. The purity of La, Fe and Si raw material are all greater than or equal to 99.9%. In order to ensure well heat distribution of these sample, button-shaped sample were smelted four times and each time they were turned over. The arc-melted ingots wrapped by Ta foils were sealed in a quartz tube filled with high-purity argon gas, then homogenized at 1050℃ for 20 days, and finally quenched into cold water.

The $LaFe_{11.5}Si_{1.5}$ particles were pulverized in an agate mortar and mixed with 20% Cu powder. Then the sample were sintered spark plasma sintering(SPS) by furnace (Syntex Lab; SPS Syntex Inc.) at 700℃ and 50MPa pressure under vacuum and pressure was held for 5 minutes. The photo of $LaFe_{11.5}Si_{1.5}$/Cu composites are shown in figure 1. (a).

2.2. *NTE and mechanical property measurement*

The linear thermal expansion data($\Delta L/L$) were measured using a strain gauge over a temperature range of 77-300K. During the process, the Zerodur glass ceramic, whose

IOP Conf. Series: Journal of Physics: Conf. Series **897** (2017) 012003 doi:10.1088/1742-6596/897/1/012003

coefficient of thermal expansion is near zero, was used for reference. The compressive strength and modulus of elasticity were measured by MTS (CMT 5000) universal material testing machine. Two samples, whose size is 3.2×3.61×10 mm^3 (1# sample) and 2.9×3.61×10 mm^3 (2# sample) respectively, were tested at 77 K and 300 K. Hardness of the material was tested by automatic rotating Vickers hardness tester.

2.3. Tensile test with carbon fibre plate

The tested sample in tensile test was carbon fibre plate with the size of 155×7×1.8 mm^3. Two pieces of Al sheet were adhered on both side of the carbon fibre plate in order to sustain the tested sample, as is shown in Figure 1. (b). $LaFe_{11.5}Si_{1.5}$/Cu NTE materials, were adhered on the upper side of one Cu sheet. The size of the two Al sheets and $LaFe_{11.5}Si_{1.5}$/Cu NTE materials is 17×7×3 mm^3 and 17×7×1.5 mm^3, respectively. In the control experiment, the places of NTE materials were substituted by Cu sheets. D-W glue was used to stick these sheets and plate together. The adhesive order is shown in Figure 1. (c).

MTS-SANS (CMT 5105) universal testing machine with a load capacity of 10kN was used in the tensile test. During tensile test at 77K, the samples and fixtures were immersed in liquid nitrogen. Tensile tests were conducted in displacement control with a stain rate of 10×10^{-4} s^{-1}. Samples were fixed by 304 LN stainless steel homemade clamp.

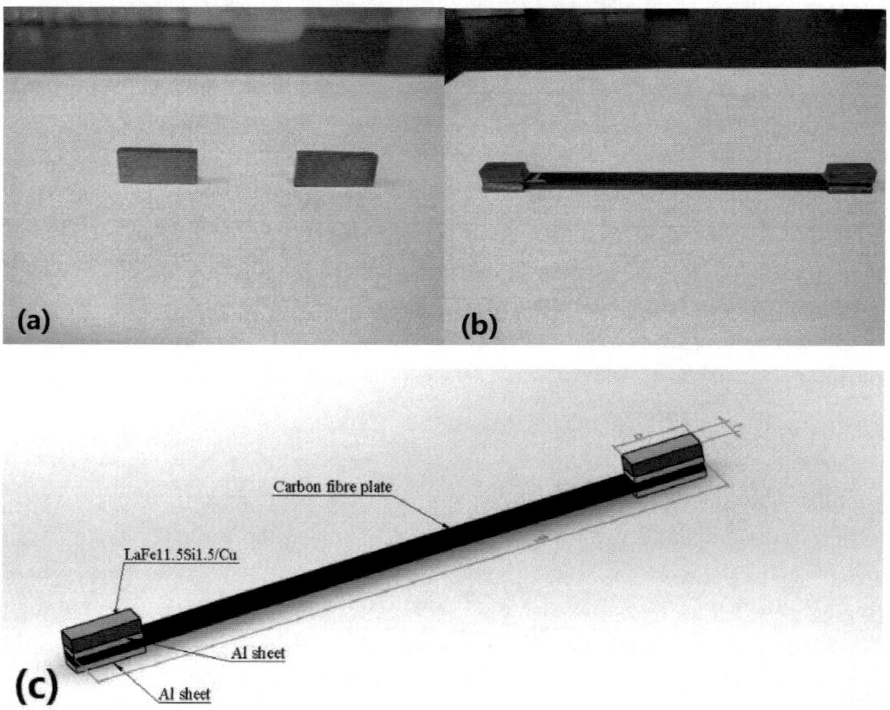

Figure 1. Photos and diagrams of the tested samples

3. Results and discussion

3.1. Abnormal thermal expansion properties

Figure 2. shows the linear thermal expansion ΔL/L data (reference temperature: 300 K) as a function of temperature for $LaFe_{11.5}Si_{1.5}$/Cu composites in the temperature range from 77 K to 300 K. It can be found that the material presents positive thermal expansion behavior(PTE) from 200 K to 300 K and shows evident NTE behavior in the temperature range of 160K to 215K. It can be calculated that the CTE of the composite is -40.06×10^{-6} K^{-1}. From room temperature zone to low temperature zone (<180 K), the strain value of the material is more than 1400ppm and presents remarkable NTE properties.

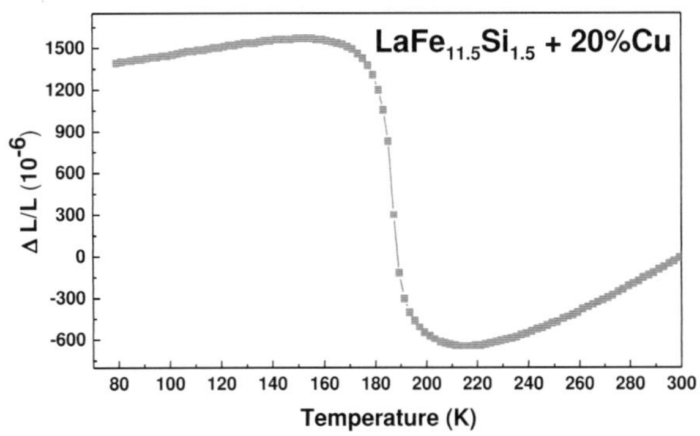

Figure 2. Temperature dependence of linear thermal expansion ΔL/L (reference temperature: 300 K) from 77K to 300K for $LaFe_{11.5}Si_{1.5}$/Cu composites.

3.2. Mechanical properties

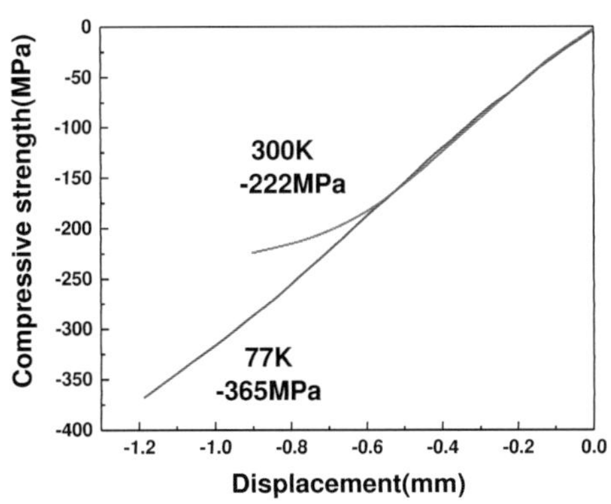

Figure 3. Compressive strength-displacement curve at 77 K and 300 K

To evaluate the compressive properties of low-temperature NTE material $LaFe_{11.5}Si_{1.5}$/Cu, compression tests were carried out at nitrogen temperature (77K) and room temperature (300K). Stress-displacement curves and stress-strain data are obtained from the tests and are used to determine the compressive strength and modulus.

Figure 3. shows the typical compression stress-displacement curves for $LaFe_{11.5}Si_{1.5}$/Cu at 77 K and 300 K, respectively. The values of compressive strength are 365MPa and 222MPa, respectively. It is found that the compressive strength of the material is much better at cryogenic temperatures than that at room temperatures, which illustrates that the material can maintain its shape to a great extent without too much deformation when compressed in cold environment.

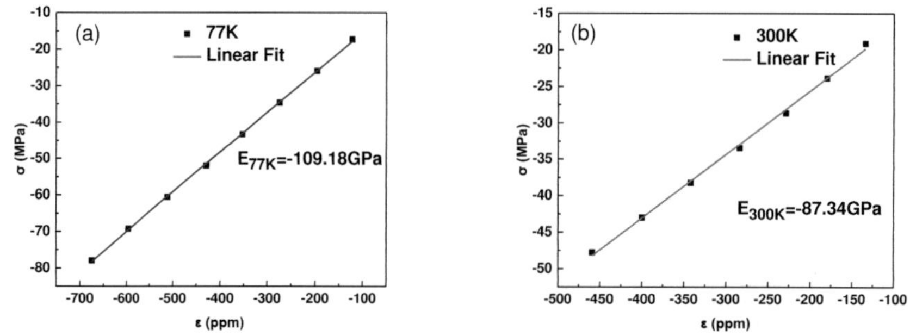

Figure 4. Stress-strain curve of the sample at 77 K and 300 K

Figure 4. shows the stress-strain data for $LaFe_{11.5}Si_{1.5}$/Cu at 77 K and 300K. From the stress-strain data, the Young's modulus at each temperature was calculated. They are -109.18 GPa and -87.34 GPa, at 77 K and 300 K, respectively. Since Young's modulus reflects the rigidity of material, the higher the absolute value is, the more impossible the material deform while traction. The absolute value of Young's modulus at 77 K is relatively higher than that at 300 K means that it has better mechanical property at cryogenic temperatures. Table 1. shows the hardness test results. The average value of Vickers hardness is 337.7 HV.

Table 1. Results of hardness test

Test Number	1	2	3	4	5	Average Value
Vickers hardness (Hv 0.5/10)	261.9	370.9	381.5	378.4	295.7	337.7

3.3. Tensile test

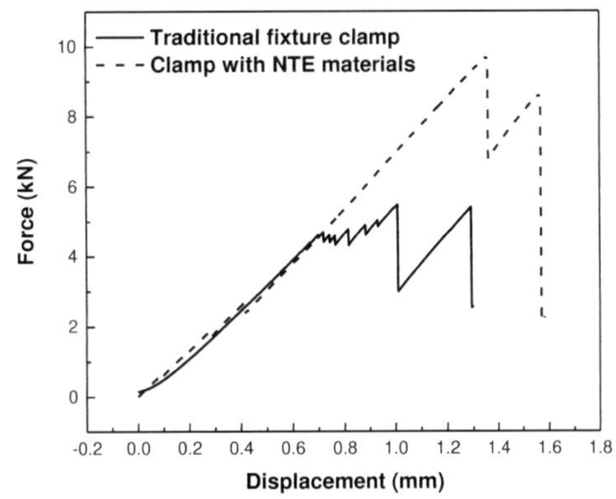

Figure 5. Force-displacement curve of carbon fibre reinforced epoxy based composite sample by traditional fixture clamp and NTE material embedded sample in the tensile test

Figure 5. clearly describes the mechanical state of the samples in the tensile test. During the test, the sample was slowly extended by the testing machine, gradually the force reached the maximum value. Then the sample slipped with the force decreasing quickly before the sample were held firmly again by the machine. Next the force began to increase. The whole process repeated several times until the sample finally pulled off by the machine. From Table 2., the results of the tensile test clearly show that, after adding $LaFe_{11.5}Si_{1.5}$/Cu NTE material on both side of the carbon fibre plate, the max force is 9.77 kN, which is about 78% more than that of traditional fixture clamp. Also, the tensile strength of carbon fibre with NTE material embedded is 775.28 MPa, which is much more than that of carbon fibre plate. From Figure 5., two curves are nearly the same at the beginning, when displacement just exceeds about 0.7mm, the curve of sample without NTE materials sheets embedded starts to slip, which means that the relative slip between the clamp and the sample occurred. By contrast, samples with NTE material embedded was held much firmly.

Table 2. Results of the tensile test

	Max Force(kN)	Tensile Strength(MPa)
Carbon fibre only	5.48	434.8
NTE material embedded	9.77	775.28

The tensile test validated that, the adding of NTE material do make carbon fibre plate held tighter by the clamps in cryogenic environment, for the NTE material's volume expands at low temperatures, which compensate the space due to the contraction of carbon fibre plate with the temperature decreasing.

4. Conclusion

In this work, $LaFe_{11.5}Si_{1.5}$/Cu NTE materials were successfully synthesized. Their NTE properties and mechanical properties were tested and analyzed. The results show that their mechanical performance, elastic modulus and hardness, is excellent at cryogenic temperatures.

Then these NTE materials were stuck on carbon fibre plate in tensile test. The results show that the expansion of NTE materials at cryogenic environment can effectively fix the tested carbon fibre plate samples. The applications of NTE materials in tensile test and other mechanical property supply simple and useful solutions to problem brought by materials' contraction at low temperatures.

Acknowledgements
This work was supported by the National Natural Science Foundation of China (No. 51522705, 51577185 and 51401224), the National Magnetic Confinement Fusion Science Program (Grant No.2015GB121001), Beijing Nova program (Grant No. xx2016072), the Foundation of CAS (Grant No. QYZDB-SSW-JSC042).

References

[1] Guillaume, C.E., *Comptes Rendus de l'Académie des Sciences*, 1897. **125**: p. 235-238.

[2] Hummel, F.A., Thermal Expansion Properties of Some Synthetic Lithia Minerals. *Journal of the American Ceramic Society*, 1951. **34**(8): p. 235-239.

[3] Mary, T.A., et al., Negative Thermal Expansion from 0.3 to 1050 Kelvin in ZrW2O8. *Science*, 1996. **272**(5258): p. 90-92.

[4] Perottoni, C.A., et al., Pressure-Induced Amorphization and Negative Thermal Expansion in ZrW2O8. *Science*, 1998. **280**(5365): p. 886-889.

[5] Chen, J., et al., The Role of spontaneous polarization in the negative thermal expansion of tetragonal PbTiO3-Based compounds. *Journal of the American Chemical Society*, 2011. **133**(29): p. 11114-11117.

[6] Li, C.W., et al., Structural Relationship between Negative Thermal Expansion and Quartic Anharmonicity of Cubic ScF_{3}. *Physical Review Letters*, 2011. **107**(19).

[7] Greve, B.K., et al., Pronounced Negative Thermal Expansion from a Simple Structure: Cubic ScF3. *Journal of the American Chemical Society*, 2010. **132**(44): p. 15496-15498.

[8] Takenaka, K. and H. Takagi, Giant negative thermal expansion in Ge-doped anti-perovskite manganese nitrides. *Applied Physics Letters*, 2005. **87**(26): p. 261902.

[9] Sun, Y., et al., Lattice contraction and magnetic and electronic transport properties of Mn[sub 3]Zn[sub 1 - x]Ge[sub x]N. *Applied Physics Letters*, 2007. **91**(23): p. 231913-3.

[10] Huang, R., et al., Giant negative thermal expansion in NaZn13-type La(Fe, Si, Co)13 compounds. *J Am Chem Soc*, 2013. **135**(31): p. 11469-72.

[11] Li, W., et al., Low-temperature negative thermal expansion property of Mn doped La(Fe,Si)13 compounds. *Journal of Alloys and Compounds*, 2015. **628**(0): p. 308-310.

[12] Li, S., et al., Low-temperature negative thermal expansion behavior of LaFe11.2Al1.8−xSix compounds. *Journal of Alloys and Compounds*, 2015. **646**: p. 119-123.

[13] Zhao, Y., et al., Effect of cobalt doping on the structural, magnetic and abnormal thermal expansion properties of NaZn13-type La(Fe1-xCox)11.4Al1.6 compounds. *Physical Chemistry Chemical Physics*, 2016.

[14] Wang, W., et al., Zero thermal expansion in NaZn13-type La(Fe,Si)13 compounds. *Physical Chemistry Chemical Physics Pccp*, 2015. **17**(4): p. 2352-6.

[15] Hu, P., et al., Thermal Expansion, Ferroelectric and Magnetic Properties in (1− x) PbTiO3− x Bi (Ni1/2Ti1/2) O3. *Journal of the American Chemical Society*, 2010. **132**(6): p. 1925-1928.

[16] Nakamura, Y., et al., Mechanical Properties of Metallic Perovskite

Mn3Cu0.5Ge0.5N:High-Stiffness Isotropic Negative Thermal Expansion Material. *Journal of the American Ceramic Society*, 2009. **92**(12): p. 2999-3003.

Properties of radiation stable insulation composites for fusion magnet

Zhixiong Wu, Rongjin Huang, Chuanjun Huang, Laifeng Li*

State Key Laboratory of Technologies in Space Cryogenic Propellants, Technical Institute of Physics and Chemistry, Chinese Academy of Sciences, Beijing 100190, China

E-mail: lfli@mail.ipc.ac.cn

Abstract. High field superconducting magnets made of Nb_3Al will be a suitable candidate for future fusion device which can provide magnetic field over 15T without critical current degradation caused by strain. The higher magnetic field and the larger current will produce a huge electromagnetic force. Therefore, it is necessary to develop high strength cryogenic structural materials and electrical insulation materials with excellent performance. On the other hand, superconducting magnets in fusion devices will experience significant nuclear radiation exposure during service. While typical structural materials like stainless steel and titanium have proven their ability to withstand these conditions, electrical insulation materials used in these coils have not fared as well. In fact, recent investigations have shown that electrical insulation breakdown is a limiting factor in the performance of high field magnets. The insulation materials used in the high field fusion magnets should be characterized by excellent mechanical properties, high radiation resistivity and good thermal conductivity. To meet these objectives, we designed various insulation materials based on epoxy resins and cyanate ester resins and investigated their processing characteristic and mechanical properties before and after irradiation at low temperature. In this paper, the recent progress of the radiation stable insulation composites for high field fusion magnet is presented. The materials have been irradiated by ^{60}Co γ-ray irradiation in air at ambient temperature with a dose rate of 300 Gy/min. The total doses of 1 MGy, 5 MGy and 10 MGy were selected to the test specimens.

1. Introduction

Compared to International Thermonuclear Experimental Reactor (ITER), the next fusion device like DEMO will require the higher magnetic field and larger plasma volume. It means that larger superconducting magnets with higher magnetic field would be designed and the conductor would endure the higher electro-magnetic force and higher levels of radiation. For example, the peak field for ITER TF is 11.8 T and 13.6 T for DEMO[1]. On the other hand, superconducting magnets in fusion devices will experience significant nuclear radiation exposure during service. Therefore, it should be noted that the irradiation effect of the superconducting magnet materials would be the main key issues for the future fusion device. Among the superconducting magnet materials, the insulation material is reported as the most radiation-sensitive, which consists of glass fiber reinforced polymer (GFRP) composites [2-4]. Generally speaking, radiation-resistant property of the composite is related to the chemical structures [5]. For example, aromatic structures are superior to aliphatic structures and radiation tolerance increases with resin functionality. Based on this, we designed various insulation

IOP Conf. Series: Journal of Physics: Conf. Series **897** (2017) 012004 doi:10.1088/1742-6596/897/1/012004

materials based on epoxy resins with different functionality and cyanate ester resins and investigated their processing characteristic and mechanical properties before and after irradiation at low temperature [6-8].

In this paper, the recent progress of the radiation stable insulation composites for fusion magnet is presented. All materials have been irradiated by ^{60}Co γ -ray irradiation in air at ambient temperature. The total doses of 1 MGy, 5 MGy and 10 MGy were selected to the test specimens.

2. Experiment

2.1. Materials

The epoxy resins used in this study were diglycidyl ether of bisphenol F epoxy resin (DGEBF, GY285, Huntsman Advanced Materials) and triglycidyl-p-aminophenol (TGPAP, MY0510, Huntsman Advanced Materials), respectively. The cyanate ester was 1,1-Bis(4-cyanatophenyl)ethane (DCBE, Primaset LECy from Lonza).The curing agent was diethyl toluene diamine (DETD, HY5200, Huntsman Advanced Materials). The modifier agent was isopropylidenebisphenol bis[(2-glycidyloxy-3-n-butoxy)-1-propylether] (IPBE, PY 4122, Huntsman Advanced Materials). The reinforcement was boron-free glass fiber cloth (RW220-90, Sinoma Science and Technology Co., China) treated with a silane coupling agent.

2.2. Specimen preparation

The composites were prepared by the VPI process. Details of the VPI process were described in reference [6] and reference [8]. The resin components are given in table 1. The prepared composites were cut into the specimens for the short beam shear test with the dimensions of 24 mm× 8 mm× 4 mm (length × width × thickness), as shown in figure 1.

Table 1. Formulations of resin systems.

Sample code	DGEBF (g)	TGPAP (g)	IPBE (g)	CE (g)	DETD (g)
1	100	0	0	0	26.7
2	90	0	10	0	25.4
3	0	100	0	0	44.5
4	0	90	10	0	41.4
5	60	0	0	40	0

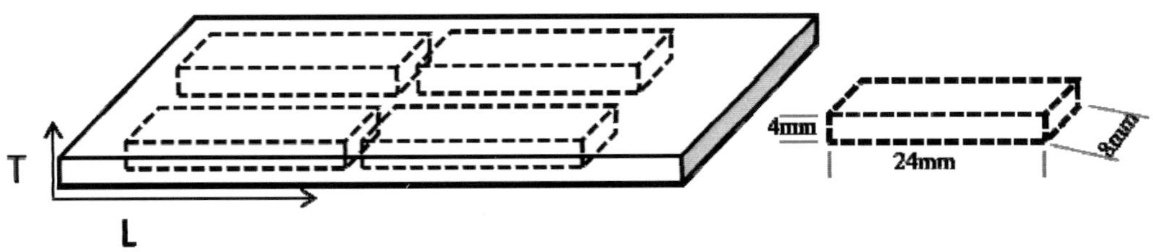

Figure 1. Schematic showing location of specimen removal from the composites.

2.3. Test procedures

The specimens were irradiated by ^{60}Co γ -ray irradiation in air at ambient temperature with a doses rate of 300 Gy/min. The maximum dose was 10 MGy.

The apparent interlaminar shear strength (ILSS) was assessed by the short-beam shear (SBS) test according to the ASTM D2344. The short beam shear (SBS) test at 77 K was carried out using a MTS-SANS 5000 universal testing with a cross-head speed of 1 mm/min and a span-to-thickness ratio of 5:1 was used. The ILSS can be calculated using the following formula:

$$ILSS = \frac{0.75P_B}{b \cdot h} \quad (1)$$

where P_B is the maximum load, b is the specimen width, and h is the specimen thickness.

3. Results and discussion

3.1. Viscosity

Generally, the insulation systems for superconducting magnets are fabricated by a vacuum press impregnation (VPI) process. The resin matrix with a rather low viscosity and long working life is essential to assure a complete impregnation of a large superconducting coil with a complicated shape [5, 9]. In this work, the resin viscosity was measured using a Brookfield viscometer. The viscosity of the resin systems at 45°C as a function of time is shown in figure 2. Clearly, compared with other resin systems, the CE/DGEBF system exhibits lower initial viscosity and longer pot life. This indicates that the CE/DGEBF system is more suitable for VPI process to impregnate large-scale superconducting magnets. Moreover, it can be seen that the addition of IPBE caused an decrease of the viscosity rise rate for both DGEBF and TGPAP. This implies that the IPBE reduced the rate of curing reactions and in turn lead to the increase of the effective working life.

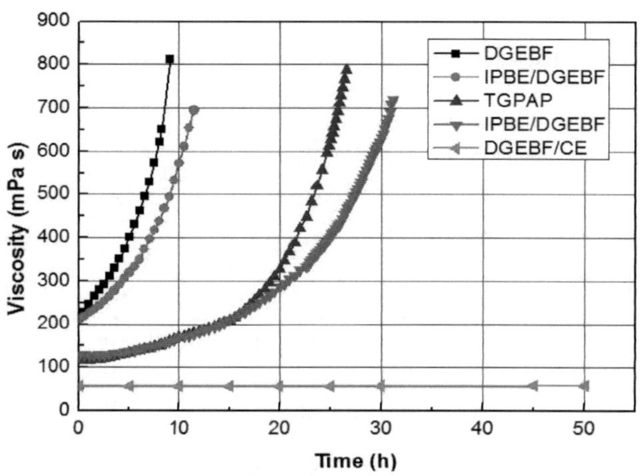

Figure 2. Viscosity versus time at 45 °C for various epoxy resin systems.

3.2. Void content

The contents of glass fiber, epoxy resin and void for the prepared composites were measured according to ASTM D2734. The results are presented in table 2. The void contents of all prepared composites were less than 1 vol.%. Furthermore, the volume contents of the fiber and the matrix for all prepared composites were about 45 vol.% and 54 vol.%, respectively.

Table 2. Volume content of glass fiber, epoxy resin and void.

Resin System #	Resin Content Vol. %	Glass Content Vol %	Void Content Vol %
DGEBF	54.47	45.29	0.24
TGPAP	53.72	45.83	0.45
IPBE/DGEBF	54.03	45.38	0.59
IPBE/TGPAP	53.84	45.84	0.32
CE/DGEBF	53.98	45.21	0.81

3.3. Interlaminar shear strength

The ILSSs of the non-irradiated and irradiated composites were measured by the SBS test at 77 K. The ILSS as a function of the irradiation dose is shown in figure 3. It can be seen that among all formulations, the IPBE/DGEBF composite presented the highest shear strength value before irradiation, while TGPAP composite showed the lowest ILSS. TGPAP is known to result in cured materials with higher crosslink density than DGEBF resin, because of the presence of three reactive groups in the resin on a single aromatic nucleus. That is to say, TGPAP is more brittle than DGEBF and thus exhibited lower shear strength at 77 K. Moreover, the addition of IPBE improved the ILSS of DGEBF and TGPAP composite. The mechanism for the improvement of ILSS was discussed in reference [10].

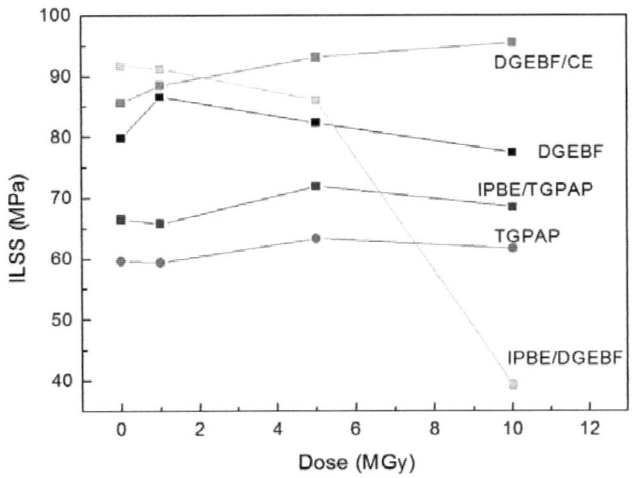

Figure 3. ILSS results for the composites, illustrating the variation with gamma-irradiation dose

Moreover, with the increasing of gamma-irradiation doses, ILSS of these composites show different trends. It is observed that ILSS of composites based on DGEBF, IPBE/TGPAP, TGPAP and IPBE/DGEBF slightly increases with the irradiation dose and then decreases. Especially for IPBE/DGEBF systems, the ILSS of composites are significantly decreased with the irradiation dose up to 10MGy. However, the DGEBF/CE systems show that the ILSS of composites increases with the increasing of irradiation doses. That is to say, IPBE/DGEBF composite could not withstand 10MGy gamma-irradiation. Moreover, DGEBF/CE systems show the best behavior under 10MGy gamma-irradiation.

SEM images of fracture surfaces of composites are shown in figure 4. Obviously, hackle pattern fracture surfaces are observed for irradiation and un-irradiation composites based on DGEBF and DGEBF/CE systems. These microstructure characteristics suggest that the interfacial adhesion of the composites after 10MGy irradiation is strong and the failure mechanisms for these composites are a combination of resin fracture and interfacial debonding. On the other hand, it is observed that fracture surface of IPBE/DGEBF composite after 10MGy irradiation shows smooth and few river lines, which indicates that the composite failure is caused by the brittle fracture of the polymer matrix. Therefore, less load was needed to destroy the composites.

The ILSS reduction mechanisms for IPBE/DGEBF composite can be related to the radiation-induced degradation in the mechanical properties of the matrix resin owing to molecular chain scission and crosslinking. In general, polymers after the high energy irradiation would be induced two competing processes: chain scission and crosslinking. IPBE has long aliphatic chain which is more easy to be destroyed by the high energy irradiation, and thus leading to a predominance of chain scission at high dose radiation, which results in the significant decrease in ILSS after 10MGy irradiation for the IPBE/DGEBF composite.

Therefore, based on the results we have discussed, DGEBF/CE composites have shown the best mechanical and irradiation resistant behaviors.

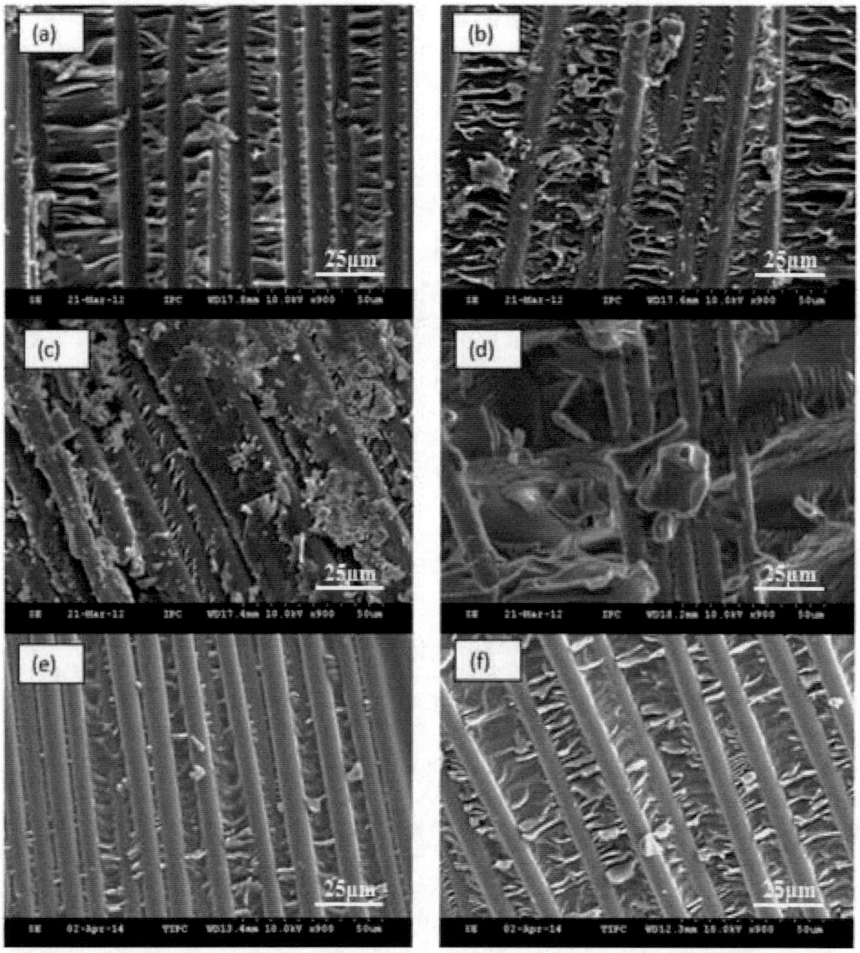

Figure 4. SEM images of fracture surfaces after SBS tests. (a) 0MGy, DGEBF; (b) 10MGy, DGEBF; (c) 0MGy, IPBE/DGEBF; (d) 10MGy, IPBE/DGEBF; (e) 0MGy, CE/DGEBF; (f) 10MGy, CE/DGEBF.

4. Conclusion

We designed various insulation materials based on epoxy resins and cyanate ester resins and investigated their processing characteristic and mechanical properties before and after irradiation at low temperature. The cyanate ester/epoxy system exhibits the lowest initial viscosity and the longest pot life which shows that the cyanate ester/epoxy system is suitable for VPI process to impregnate large-scale superconducting magnets. All insulation materials except for IPBE/DGEBF show good radiation resistance, especially CE/DGEBF. From these limited results, the CE/DGEBF system appears to be more suitable for the high field fusion magnets.

Acknowledgments

This work was supported by the National Natural Science Foundation of China (Grant No.: 51407180, 51577185).

References

[1] Bruzzone P 2014 *Fusion Eng. Des.* 89 1775

[2] Matsui K, Hemmi T, Kajitani H, Yamane M, Mizutani T, Nakano T, Ando S and Koizumi N 2016 *Fusion Eng. Des.* 109–111 574

[3] Nishimura A, Izumi Y, Imaizumi M, Nishijima S, Hemmi T and Shikama T 2011 *Fusion Eng. Des.* 86 1558

[4] Reed R, Fabian P, Bauer-McDaniel T, Hazelton C and Munshi N 1995 *Cryogenics* 35 739

[5] Evans D and Canfer S. 2000 *Adv. Cryog. Eng.* 46A 361-368.

[6] Wu Z, Li J, Huang C, Huang R and Li L 2013 *J. Nucl. Mater.* 441 67

[7] Wu Z, Li J, Huang C, Huang R and Li L. 2013 *Fusion Eng. Des.* 88 3078

[8] Li J, Wu Z, Huang C and Li L. 2014 *Fusion Eng. Des.* 89 3112

[9] Hemmi T, Koizumi N, Matsui K, Okuno K, Nishimura A, Sakai M and Asano S 2009 *Fusion Eng. Des.* 84 923

[10] Wu Z, Li J, Huang C and Li L 2015 *Physics Procedia* 67 1068

Low-temperature abnormal thermal expansion property of Mn doped cubic $NaZn_{13}$-type $La(Fe, Al)_{13}$ compounds

Yuqiang Zhao[1,2], Rongjin Huang[1,2,*], Yi Shan[1,2], Wei Wang[1], Jiangtao Li[1,2] and Laifeng Li[1,2,*]

[1]State Key Laboratory of Technologies in Space Cryogenic Propellants, Technical Institute of Physics and Chemistry, Chinese Academy of Sciences, Beijing 100190, P. R. China.

[2]University of Chinese Academy of Sciences, Beijing 100049, P. R. China

[*]E-mails: huangrongjin@mail.ipc.ac.cn; laifengli@mail.ipc.ac.cn

Abstract. Low-temperature abnormal thermal expansion (ATE) materials have been recently developed because of their significant applications for cryogenic engineering. However, the challenge still remains for the control of ATE effect at cryogenic temperature and adjustable ATE is of fundamental interest. In this paper, we report the isotropic ATE in $La(Fe, Al)_{13}$ compounds over a wide adjusting temperature range by partially substituting Fe by Mn. It is found that all samples crystallize in the cubic $NaZn_{13}$-type structure with the $Fm\overline{3}c$ space group. The introduction of nonmagnetic Mn atoms reduces the Fe-Fe exchange interaction, therefore, the itinerant electron system needs less energy to break the magnetic order in ferromagnetic (FM) state at low temperature. The negative thermal expansion (NTE) operation-temperature window moves towards lower temperatures accompanied with the decrease of Curie temperature (T_C) by increasing Mn elements. Moreover, the composite combining Mn 0 and Mn 57 broadens the zero thermal expansion (ZTE) behavior occurring in the whole tested temperature range. The present studies could be useful to control the thermal expansion, and indicate the potential applications of ATE materials in cryogenic engineering.

1. Introduction

Most materials expand when heated and contract on cooling due to the effect of the anharmonic lattice vibration. The thermal induced strain usually gives rise to materials' physics and function failure originated from the undesirable mismatch of the coefficient of thermal expansion (CTE) in different components[1] and the decline in the precision of thermometric instruments[2]. Therefore the abnormal thermal expansion (ATE) property including zero thermal expansion (ZTE) and negative thermal expansion (NTE) has become an advanced research focus especially in high-precision devices. In recent years, several kinds of materials with ATE property have been discovered, such as $LiAlSiO_4$ (β-eucryptite), ZrW_2O_8,[3] ScF_3,[4, 5] $PbTio_3$-based compounds,[6] $La(Fe, Si)_{13}$-based compounds[7-10] and some antiperovskite manganese nitrides[2, 11]. It is noteworthy that most of ATE materials are tuned to adjust the temperature range of ATE moving towards higher temperature span. However, the low-temperature volume control across a wide temperature range is also one of the most important functions and has high potential applications in the rapid developing cryogenic engineering, such as low-temperature switch, cryogenic valve, piston/piston sleeve of refrigerator and carrying structure of superconducting magnet.[12]

La(Fe, M)$_{13}$-based (M = Si, Al) rare earth intermetallic compounds recently have attracted more considerable attention as the candidates for ATE materials due to their isotropic and controllable ATE properties, simple preparation technology, high electrical and thermal conductivity, and good mechanical performance.[7] The NTE property at low temperature is required to counteract the normal positive thermal expansion (PTE). However, the starting NTE operation-temperature window in ferromagnetic (FM) La(Fe, M)$_{13}$-based compounds is generally above 100 K, which is too high for practical applications in cryogenic engineering.[7, 9] Moreover, nearly ZTE behavior could be observed in some of the materials while most of them show relatively narrow ZTE temperature ranges.[13, 14] Nevertheless, it has been reported that the $LaFe_{11.4}Al_{1.6}$ compound shows a slight NTE property in a temperature range 150 - 255 K, and exhibits the ZTE behavior in temperature ranges of 5 - 150 K and 255 - 300 K.[15] Here, we report an effective way to control the ATE behavior shifting towards the lower temperature region and occurring over a large scale by optimizing the chemical composition. We denote the $LaFe_{11.4}Al_{1.6}$ compound without Mn dopant as Mn 0. According to the structure analysis, the Mn atoms enter the 96i site substituting for Fe^{II} randomly, and the Mn occupancy at the site is calculated to be 3.8%, 5.7% and 7.6% for the three prepared samples. Correspondingly, the compounds with the increase of Mn content are denoted as Mn 38, Mn 57 and Mn 76, respectively.

In this letter, the effect of Mn doping on the structural, magnetic and abnormal thermal expansion properties of the $LaFe_{11.4}Al_{1.6}$ compound were extensively explored and discussed. All samples have the same cubic $NaZn_{13}$-type structure (space group: $Fm\bar{3}c$). It was found that the NTE temperature window shifts towards the lower temperature region with the increase of Mn concentration. Furthermore, the NTE material Mn 57 can combine the Mn 0 to exhibit ZTE behavior in the temperature range of 150 - 255 K, thus the ZTE operation-temperature window can be broadened to the whole tested temperature range.

2. Experimental procedure

We prepared the polycrystalline samples of Mn 0, Mn 38, Mn 57 and Mn 76 in a vacuum arc melting furnace under a high-purity argon atmosphere. The raw materials of Fe, Mn, Al, and La are all at least 99.9 wt % pure. An excess 10 % of La over the stoichiometric composition was added to compensate for the loss during melting. Button-like samples were melted four times, and each time buttons were turned over during melting to ensure homogeneity in composition. Arc-melted ingots were sealed in vacuum quartz tubes, followed by quenching at 950 °C for 10 days, and finally cooled down quickly to room temperature.

The linear thermal expansion data ($\Delta L/L$) were measured using a strain gage (gage factor: 1.74, type: BB120-2AA250) over the temperature range of 5 - 300 K. We have analyzed the crystal structure of the samples using X-ray diffraction (XRD) on a Bruker D8 advanced X-ray diffractometer with Cu K_α radiation ($K_{\alpha1}$ = 1.5406 Å and $K_{\alpha2}$ = 1.5443 Å). The angular scanning range was set to 20° to 80° with a step of 0.005° and scanning rate 0.5 s/step. The Fe^{I}-Fe^{II} bond length was calculated by the first-principles geometry optimization calculation using CASTEP package. Measurements of magnetization at different temperatures were performed on the physical property measurement system (Quantum Design).

3. Results and discussion

3.1. Abnormal thermal expansion

The effect of Mn doping on abnormal thermal expansion (ATE) properties of the $LaFe_{11.4}Al_{1.6}$ compound was extensively explored based on their macro-strain. Figure 1(a) displays the linear thermal expansion $\Delta L/L$ data (reference temperature: 300 K) as a function of temperature for Mn 0, Mn 38, Mn 57 and Mn 76 in the temperature range from 5 K to 300 K. It has been reported that the Mn 0 material undergoes a slight increase with decreasing temperature from about 255 K to 150 K, and shows an excellent zero thermal expansion (ZTE) behavior below 150 K.[15] However, a sharp volume change can be found at relatively low temperatures by introducing Mn atoms, which indicates that large negative thermal expansion (NTE) occurs. It is obvious that each $\Delta L/L$ versus T curve with different Mn content presents different NTE operation-temperature windows and coefficients of thermal expansion, demonstrating that the NTE property of the $LaFe_{11.4}Al_{1.6}$ compound can be effectively adjusted by optimizing Mn concentration. The coefficient of thermal expansion α (CTE) is defined as $((\Delta L/L)_{T2} - (\Delta L/L)_{T1})/(T_2 - T_1)$ from the $\Delta L/L$ versus T curves. For Mn 38, the value of the $\Delta L/L$ increases with decreasing temperature from 50 K to 175 K, indicating the NTE behavior across a temperature range of ΔT = 125 K (50 - 175 K) with a negative CTE calculated to be -11.38 $\times$ 10^{-6} K^{-1}. Furthermore, the temperature range is ΔT = 145 K (5 - 150 K) for Mn 57 while it is ΔT = 120 K (5 - 125 K) for Mn 76. Therefore, the NTE operation-temperature window decreases by increasing the amount of Mn. However, the average CTE slightly decreases as the increase of Mn content. The CTE is -6.23 $\times$ 10^{-6} K^{-1} and -4.62 $\times$ 10^{-6} K^{-1} for the samples of Mn 57 and Mn 76, respectively, which is significantly lower than that of the Mn 38 compound.

More fascinatingly, the NTE property can counteract the normal positive thermal expansion (PTE) to exhibit almost zero thermal expansion (ZTE) behavior. Figure 1(b) shows the temperature dependence of linear thermal expansion $\Delta L/L$ data for Mn 0, Mn 57 and a mixture of Mn 0 and Mn 57 with equal amount. It was found that the ZTE can occur in the span of 150 - 255 K by combining the Mn 0 and Mn 57 materials. The origination of ZTE for the composite is specifically explained in the inset of Figure 1(b). In the certain temperature range 150 - 255 K, the Mn 0 compound shows a thermal expanding while Mn 57 gives rise to a thermal contraction due to the temperature decrease. Importantly, the volume expansion just compensates the volume contraction, thus the ZTE behavior can be observed and modified for a wider operation-temperature range.

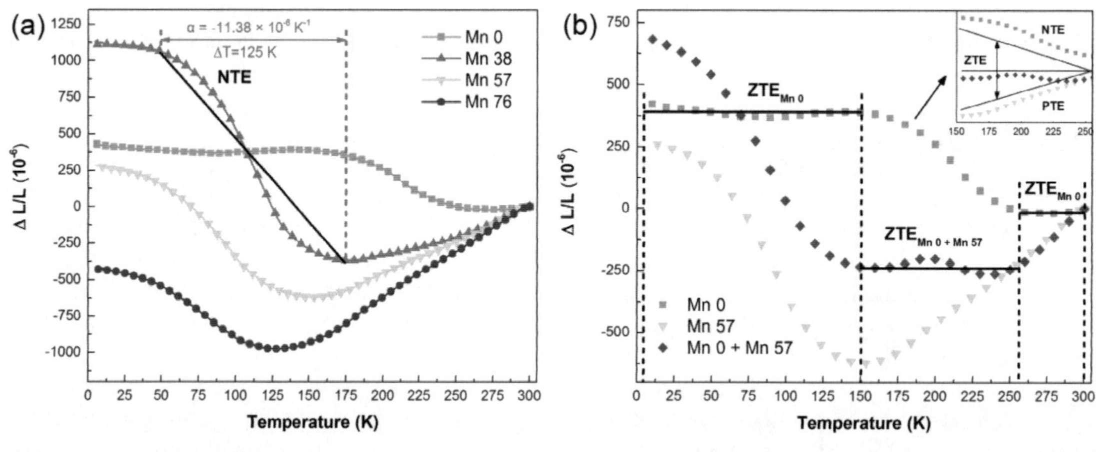

Figure 1. (a) Temperature dependence of linear thermal expansion $\Delta L/L$ (reference temperature: 300 K) from 5 K to 300 K for Mn 0, Mn 38, Mn 57 and Mn 76 compounds. (b) Temperature dependence of linear thermal expansion $\Delta L/L$ data for the mixture of Mn 0 and Mn 57 with equal amount. The inset shows the ZTE behavior of the composite from 150 K to 255 K.

3.2. X-ray diffraction and crystal structure

Previous research has revealed that the Mn 0 compound crystallizes in a nearly single phase of the cubic $NaZn_{13}$-type structure (space group: $Fm\bar{3}c$).[15] To explore the effect of Mn dopant on the crystal structure, the powder X-ray diffraction (XRD) patterns were obtained at room temperature, as shown in Figure 2(a). It can be found that the crystal structure of Mn 38, Mn 57 and Mn 76 remains cubic $NaZn_{13}$-type structure based on the dominant diffraction peaks of the 1:13 phase. In the inset of Figure 2(a), it can be seen that the lattice constant gradually decreases by the further substitution of Mn for Fe, in accordance with the smaller atomic radius of Mn than that of Fe. Furthermore, a small quantity of the

α-Fe phase can be observed calculated to be less than 0.93 wt % and has a negligibly minor effect on the overall ATE property.

The $NaZn_{13}$-type crystal structure of the Mn doped $La(Fe, Al)_{13}$ compounds was shown in Figure 2(b). The Fe atoms are divided into Fe^{I} (8b) and Fe^{II} (96i) according to their different locations in the lattice. Moreover, Mn atoms only enter the 96i sites substituting Fe^{II} randomly. Therefore, the sites of 8a, 8b and 96i are occupied by the atoms of La, Fe^{I} and (Fe^{II}/Al/Mn), respectively. Here, the atom of Fe^{I} has an fcc-like local environment surrounded by 12 nearest neighbors, and the interatomic plays a significant role in the magnetic exchange interaction. In other words, when the distance of the Fe^{I}-Fe^{II} pair is smaller than 2.45 Å, the exchange interaction is negative while the interaction is positive at larger Fe^{I}-Fe^{II} distances. For $La(Fe_{1-x}Mn_x)_{11.4}Al_{1.6}$ compounds, the average Fe^{I}-Fe^{II} distance is 2.38186 Å, 2.38314 Å and 2.39678 Å for Mn 38, Mn 57 and Mn 76, respectively. Hence, the decrease of the NTE operation-temperature window can be explained by the substitution of Mn for Fe, which leads to the increase of the Fe^{I}-Fe^{II} distance and decreases the exchange interaction.

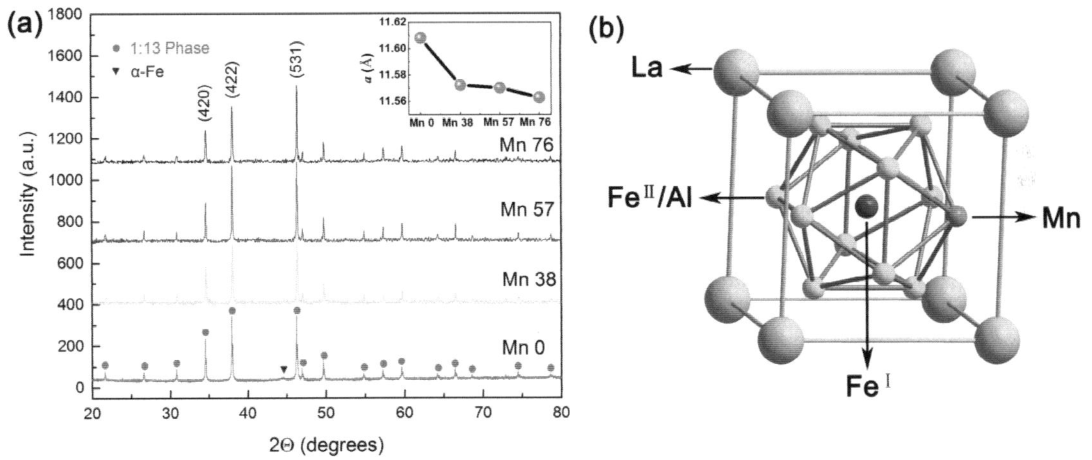

Figure 2. (a) XRD patterns for Mn 0, Mn 38, Mn 57 and Mn 76 compounds. The inset shows the lattice constant for the compounds. (b) $NaZn_{13}$-type structure (space group: $Fm\overline{3}c$) of the compounds with partial substitution of Mn for Fe.

3.3. Magnetic properties and mechanism study

Figure 3(a) presents the temperature dependence (5 - 300 K) of the magnetization $M(T)$ for all the samples measured in a magnetic field of 0.1 T. It was reported that the Mn 0 compound shows a second-order transition from the antiferromagnetic (AFM) to the paramagnetic (PM) state as the temperature reach the Neel temperature T_N.[15] However, distinctly different with Mn 0, a sharp magnetization change can be observed with the increase of temperature for Mn 38, Mn 57 and Mn 76, indicating that a ferromagnetic (FM) to PM transition occurs. The Curie temperature (T_C) was determined with the smallest rates of change of the magnetization (dM/dT) value from dM/dT curves, as shown in Figure 3(b). It is obvious that the T_C moves towards the lower temperature with more Mn atoms inserted into the lattice. Compared with the NTE property shown in Figure 1, we found that the NTE just occurs around the T_C, which is 140 K, 110 K and 90 K for Mn 38, Mn 57 and Mn 76, respectively. Therefore the decrease of the T_C makes the NTE behavior moving to low temperatures due to the weak coupling between magnetic transitions. Moreover, the decrease of average CTE as the increase of Mn content may be due to the decrease of the intensity of magnetization in the low-temperature FM state. Such optimization makes these ATE materials to be promising candidates for broader application in low temperature sensitive equipment and devices.

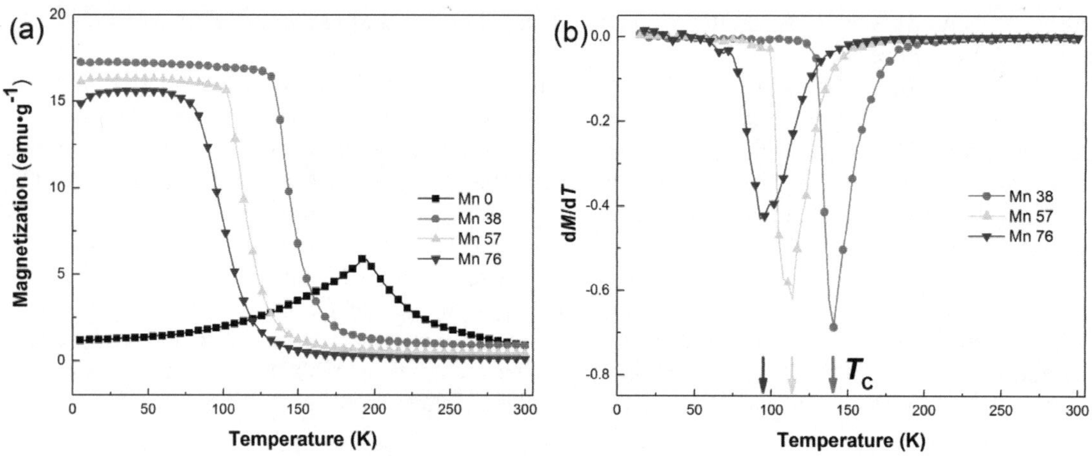

Figure 3. (a) Temperature dependence (5 - 300 K) of magnetization $M(T)$ for Mn 0, Mn 38, Mn 57 and Mn 76 compounds measured under a field of 0.1 T. (b) The rate of change of magnetization as a function of temperature for Mn 38, Mn 57 and Mn 76 compounds.

4. Conclusion

In summary, we have investigated the effect of Mn doping on ATE property for $NaZn_{13}$-type Mn 0, Mn 38, Mn 57 and Mn 76 compounds. The results reveal that the NTE operation-temperature window decreases from $\Delta T = 105$ K (150 - 255 K) to $\Delta T = 120$ K (5 - 125 K) by introducing Mn atoms, and such a low temperature window with the NTE property makes the Mn doped samples suitable for practical applications in low temperature sensitive equipment and devices. Furthermore, the composite combining Mn 0 and Mn 57 broadens the ZTE behavior occurring in the whole tested temperature range. Magnetic measurements show that the ATE property is mainly correlated with the magnetic phase transition. The ATE discovery of Mn doped La(Fe, $Al)_{13}$ compounds at low temperature would improve their potential applications in cryogenic engineering.

Acknowledgments

This work was supported by the National Natural Science Foundation of China (No. 51522705, 51577185, 51427806, 51401224 and 51408586), the National Magnetic Confinement Fusion Science Program (Grant No. 2015GB121001), Beijing Nova program (Grant No. Z161100004916114), and the Foundation of CAS (Grant No. QYZDB-SSW-JSC042).

References

[1] Chen J, Wang FF, Huang QZ, Hu L, Song XP, Deng JX, Yu RB and Xing XR 2013 *Sci. Rep.* **3** 2458.
[2] Song XY, Sun ZH, Huang QZ, Rettenmayr M, Liu XM, Seyring M, Li GN, Rao GH and Yin FX 2011 *Adv. Mater.* **23** 4690.
[3] Mary TA, Evans JSO, Vogt T and Sleight AW 1996 *Science* **272** 90.
[4] Li CW, Tang XL, Munoz JA, Keith JB, Tracy SJ, Abernathy DL and Fultz B 2011 *Phys. Rev. Lett.* **107** 195504.
[5] Greve BK, Martin KL, Lee PL, Chupas PJ, Chapman KW and Wilkinson AP 2010 *J. Am. Chem. Soc.* **132** 15496.
[6] Chen J, Xing XR, Sun C, Hu PG, Yu RB, Wang XW and Li LH 2008 *J. Am. Chem. Soc.* **130** 1144.
[7] Huang RJ, Liu YY, Fan W, Tan J, Xiao FR, Qian LH and Li LF 2013 *J. Am. Chem. Soc.* **135** 11469.
[8] Li W, Huang RJ, Wang W, Tan J, Zhao YQ, Li SP, Huang CJ, Shen J and Li LF 2014 *Inorg. Chem.* **53** 5869.

IOP Conf. Series: Journal of Physics: Conf. Series **897** (2017) 012005 doi:10.1088/1742-6596/897/1/012005

[9] Wang W, Huang RJ, Li W, Tan J, Zhao YQ, Li SP, Huang CJ and Li LF 2015 *Phys. Chem. Chem. Phys.* **17** 2352.
[10] Li SP, Huang RJ, Zhao YQ, Li W, Wang W, Huang CJ, Gong PF, Lin ZS and Li LF 2015 *Inorg. Chem.* **54** 7868.
[11] Takenaka K and Takagi H 2009 *Appl. Phys. Lett.* **94** 131904.
[12] Li W, Huang RJ, Wang W, Liu HM, Han YM, Huang CJ and Li LF 2015 *J. Alloys. Compd.* **628** 308.
[13] Li W, Huang RJ, Wang W, Zhao YQ, Li SP, Huang CJ and Li LF 2015 *Phys. Chem. Chem. Phys.* **17** 5556.
[14] Li SP, Huang RJ, Zhao YQ, Wang W and Li LF 2015 *Phys. Chem. Chem. Phys.* **17** 30999.
[15] Zhao YQ, Huang RJ, Li SP, Wang W, Jiang XX, Lin ZS, Li JT and Li LF 2016 *Phys. Chem. Chem. Phys.* **18** 20276.

Alternating current breakdown voltage of ice electret

Y Oshika[1], Y Tsuchiya[1], T Okumura[2], and Y Muramoto[1]

[1] Department of Electrical and Electronic Engineering, Meijo University,1-501 Shiogamaguchi, Tempaku-ku, Nagoya, 468-8502, Japan

[2] Department of Electrical, Electronic, and Communication Engineering, Iwate University, Ueda 4-3-5, Morioka 020-8551, Iwate, Japan

E-mail: 143441503@ccalumni.meijo-u.ac.jp

Abstract. Ice has low environmental impact. Our research objectives are to study the availability of ice as a dielectric insulating material at cryogenic temperatures. We focus on ferroelectric ice (iceXI) at cryogenic temperatures. The properties of iceXI, including its formation, are not clear. We attempted to obtain the polarized ice that was similar to iceXI under the applied voltage and cooling to 77 K. The polarized ice have a wide range of engineering applications as electronic materials at cryogenic temperatures. This polarized ice is called ice electret. The structural difference between ice electret and normal ice is only the positions of protons. The effects of the proton arrangement on the breakdown voltage of ice electret were shown because electrical properties are influenced by the structure of ice. We observed an alternating current (ac) breakdown voltage of ice electret and normal ice at 77 K. The mean and minimum ac breakdown voltage values of ice electret were higher than those of normal ice. We considered that the electrically weak part of the normal ice was improved by applied a direct electric field.

1. Introduction

Ice has low environmental impact. The authors have been studying the behavior of ice as dielectrics or electrical insulating materials at cryogenic temperatures [1–5]. We focus on the dielectric properties of ferroelectric ice (iceXI) and normal ice (iceIh) at cryogenic temperatures. The properties of iceXI, including its formation procedure, are unclear and are under investigation [6–8]. The formation procedure for iceXI requires either specific techniques with added impurities, or a very long time (at least 624 hours [8]). We attempted to obtain polarized ice, which is similar to iceXI, with neither added impurities nor very long time. The polarized ice have a wide range of engineering applications as electronic materials at cryogenic temperatures. We obtained the polarized ice by controlling the movement of hydrogen ions (protons) in ice [4–5].

Proton (H^+) movement depends on the temperature. Protons in normal ice can hop along lattice defects at relatively high temperatures, e.g. 253 K. The dependency of the proton movement on the temperature can be explained in terms of the dielectric constant. The relative permittivity of ice at 253 K is 45, and it is 3 at 77 K [9–10]. These results indicate that the responsible speed of a proton in an external electric field is low at 77 K and is high at 253 K. The proton movements can be controlled under an external electric field because the protons are positively charged. However, at cryogenic temperatures, such as 77 K, it is difficult to move protons in ice. Thus, the proton movements cannot be controlled by the electric field. The dipole moment of an H_2O molecule does not govern the dielectric properties of ice, but the proton movement can govern it. These characteristics are responsible for the dielectric polarization of ice.

We successfully obtained the polarized ice as follows. A direct electric field was applied to normal ice at around 253 K, and the ice was cooled to 77 K. During this procedure, the protons moved toward the cathode because of the electric field application. Since the position of the protons was fixed at 77 K, the ice polarized without the electric field. This polarized ice is called ice electret [11]. Electret is a dielectric material possessing permanent and spontaneous polarization. In our previous paper, we obtained ice electret under the direct electric field application with cooling, and confirmed that ice electret was polarized based on the observation of the depolarization current of ice electret that was warmed from 77 K [4–5]. The protons in ice electret were fixed at the cathode side under the direct electric field application and cooling. On the other hand, the protons in normal ice are randomly fixed at 77 K.

Ice electrets and normal ice differ in their proton arrangements. We assumed that they also differed in their electrical properties. In this paper, we studied the effect of proton arrangement on ac breakdown voltage. Our research objective is to explore the availability of ice electret as a dielectric insulating materials at cryogenic temperatures.

2. Experimental procedure

2.1. Sample configuration

Figure 1 shows the parallel-plane electrode system used in the experiments. The electrodes were made of stainless steel. The diameters of the anode and the cathode electrode were 25 mm and 16 mm, respectively. The guard electrode configuration was ring-shaped. The internal and external diameters of the guard electrode were 18 mm and 26 mm, respectively. The guard electrode plays the role of a restraint leakage current. Only the bulk current between the anode and cathode can be observed because the surface current is released to ground from the guard electrode. The thicknesses of these electrodes were 10 mm. Each electrode was fixed on acryl plates.

Figure 2 shows a flowchart of the sample preparation. In procedure III and N-1 to N-2, all electrodes were grounded because deviations in the charge were prevented. In procedure E-3, the extra charge excluding the polarized charge (e.g. free charge) was released owing to the grounding of electrodes. The cooling rate from 253 K to 77 K was approximately 1 K/min.

The distilled water between the electrodes was renewed, and the surfaces of the electrodes were polished and washed for every experiment. The temperature of the sample was observed by a T-thermocouple inserted into a dummy sample, to which an electric field was not applied.

2.2. Observation procedure of depolarization current of ice electret and normal ice

A depolarization current was observed while the temperature of the sample was increased from 77 K to 253 K as follows. After the procedure in Subsection 2.1, the sample was moved from LN_2 to an electromagnetic shield at 253 K in a freezer. The electrode system with a sample was connected to a current measurement system. Figure 3 shows the measurement circuit for the depolarization current. The measurement equipment was an electrometer/high resistance meter (Keysight, B2985A). The

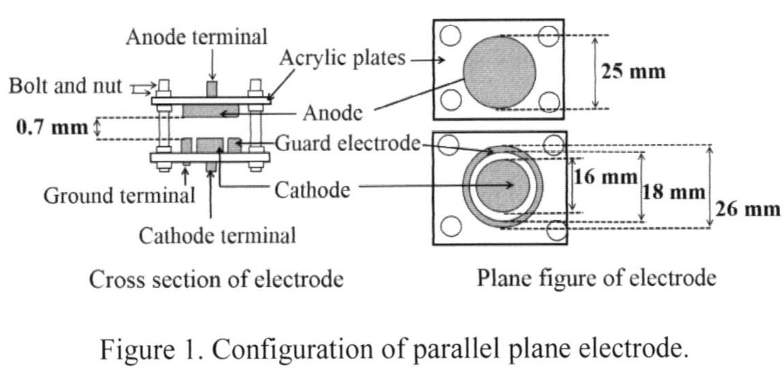

Figure 1. Configuration of parallel plane electrode.

IOP Conf. Series: Journal of Physics: Conf. Series **897** (2017) 012006 doi:10.1088/1742-6596/897/1/012006

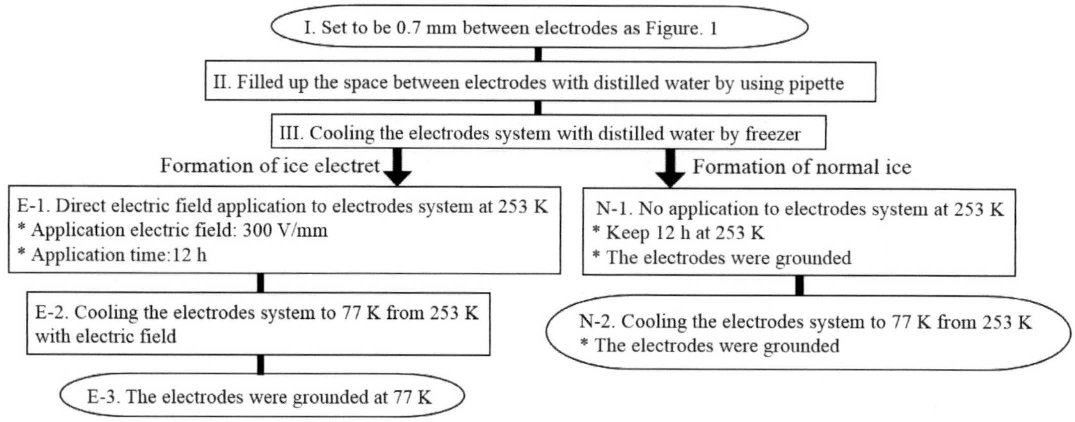

Figure 2. Procedure of sample preparation.

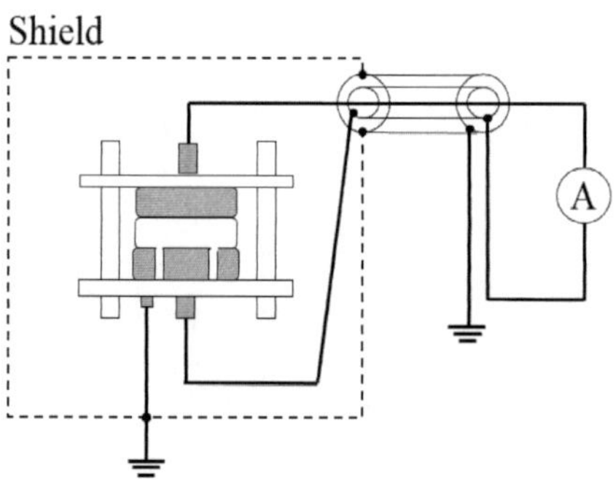

Figure 3. Measurement system of depolarization current.

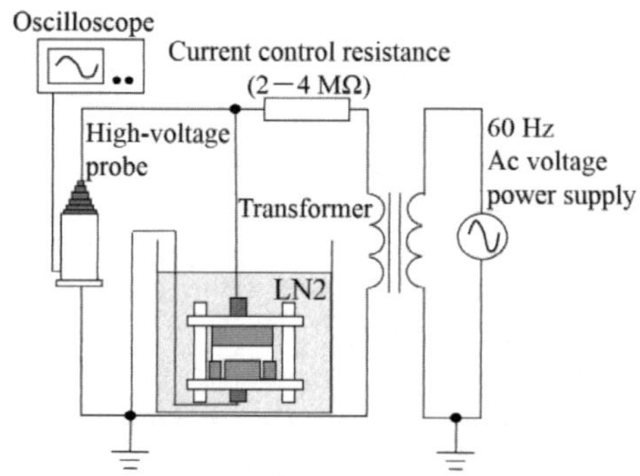

Figure 4. Diagram of breakdown experimental circuit.

depolarization current of normal ice was observed in the same manner.

2.3. Observation procedure of ac breakdown voltage

After the procedure of Subsection 2.1, the sample was placed into LN_2 in a Dewar vessel made of stainless steel. The electrodes of the sample were connected to a circuit to observe the ac breakdown voltage, as shown in figure 4. Figure 4 shows a diagram of the experimental circuit. The sample was immersed in LN_2 at 77 K and 1,013 hPa with an ac voltage application of 60 Hz to the anode electrode. The cathode electrode was grounded. The voltage application was an ac ramp voltage with a rising rate of 240 Vrms/s. Vrms is the effective value of alternating voltage ($2^{-\frac{1}{2}}$ the amplitude of voltage). Equation (1) is the ac voltage application waveform:

$$V = 240t \sin(2\pi ft) \tag{1}$$

where t is the time of the application, and f is the frequency.

In figure 4, the current control resistance prevents excessive currents in the circuit. The amplitude of the voltage applied to the sample was monitored with a high-voltage probe (EP-100K, NISSIN PULSE ELECTRONICS CO. LTD.) and an oscilloscope (DPO 4034, Tektronix). The waveform of the ac voltage was monitored from the start of the voltage application until the breakdown.

The occurrence of a breakdown of a sample can be judged by a rapid decrease in voltage. This is because once a breakdown occurs the electrodes of the sample are shortened. The amplitude of the voltage just before a breakdown is defined as a breakdown voltage. After the experiments, the electrodes were eliminated from the ice, and the position of the breakdown path was observed. A result was eliminated when the breakdown occurred through the outside of the anode and the cathode.

3. Results and Discussion

3.1. Depolarization current of ice electret and normal ice

The observation of a depolarization current indicates the presence or absence of polarization in a sample. Figure 5 shows the dependency of the depolarization current on time. The temperature was given by the dummy sample. The horizontal axis was a converted logarithmic axis. The depolarization current from 2 to 4 min was noise derived from the operation, which connected the electrode systems to measurement systems. For normal ice, positive values were observed at around 125 K and 251 K, and negative values were observed at around 168 K and 236 K in depolarization current. Because these values quickly increased and decreased repeatedly, the peak of depolarization current did not show clearly. For ice electret, the depolarization current peaked at approximately 105 K and 251 K, which was higher than that of normal ice.

We previously reported that the peak value around 105 K increased as the direct electric field increased during the ice electret procedure [5]. The relationship between the peak value at around 251 K and the electric field intensity is currently under consideration. The depolarization current peaks at around 105 K and 251 K higher than that of normal ice, and the peak around 105 K increased as the electric field increased in the procedure. These results indicate that the polarized charge fixed by protons was released at around 105 K and 251 K in ice electret.

The polarization was induced under a direct electric field application and cooling in ice electret. This result was a result of the release of protons fixed in ice electret at the cathode side. Protons began

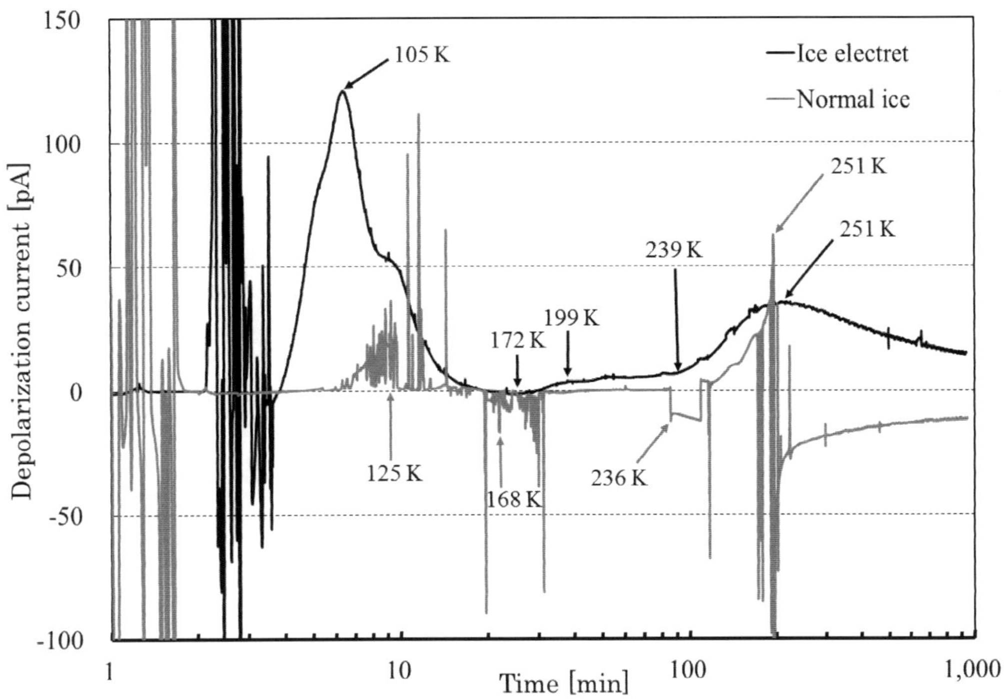

Figure 5. Dependency of depolarization current on time.

to move with thermal stimulus when the temperature increased. Some protons moved at around 105 K, and others moved at around 251 K. The reason that the depolarization current of ice electret converged at 0 pA at around 172 K is currently under investigation.

3.2. AC breakdown voltage of ice electret and normal ice

Figure 6 shows the breakdown voltage of ice electret and normal ice. The vertical axis shows the effective value of the ac breakdown voltage. Numbers in brackets are the number of samples. The maximum values of ice electret and normal ice were almost the same. The minimum value of ice electret was 30 kVrms, which was 2.5 times larger than that of normal ice (12 kVrms). For these results of normal ice (iceIh), breakdowns occurred even in the range of 12–17 kVrms. Breakdown voltages in this range must have occurred in electrically weak parts of the samples. By contrast, breakdowns did not occur below 30 kVrms for ice electret.

In these results for ice electret, the distribution of the breakdown voltage was suppressed. Increases in the mean breakdown voltage were primarily caused by improving the minimum breakdown voltage value. The measured values around 30–44 kVrms of ice electret seemed to be slightly higher than those of normal ice. There is a possibility that the electrically weak part in the normal ice was improved because breakdowns at 12–17 kVrms did not occur in ice electret. The electrically weak part of the normal ice was improved owing to the change in proton arrangement of the ice under a direct electric field. Each proton arrangement of ice electret and normal ice was maintained as observing the breakdown voltage at 77 K. The temperature dependence of depolarization current peaks in figure 5 indicated maintenance of proton arrangement.

The structural difference between normal ice and ice electret is a result only of the position of protons because the direct electric field was applied to solid H_2O. Ice electret and normal ice have the same oxygen distribution. It is considered that the arrangement of protons in the ice electret is uniform because the characteristics of the proton arrangement are uniform in iceXI, which is similar to ice electret that holds its polarization at cryogenic temperatures. The positions of oxygen atoms in iceIh (normal ice) and iceXI are almost the same, and have a hexagonal arrangement. Figure 7 shows a

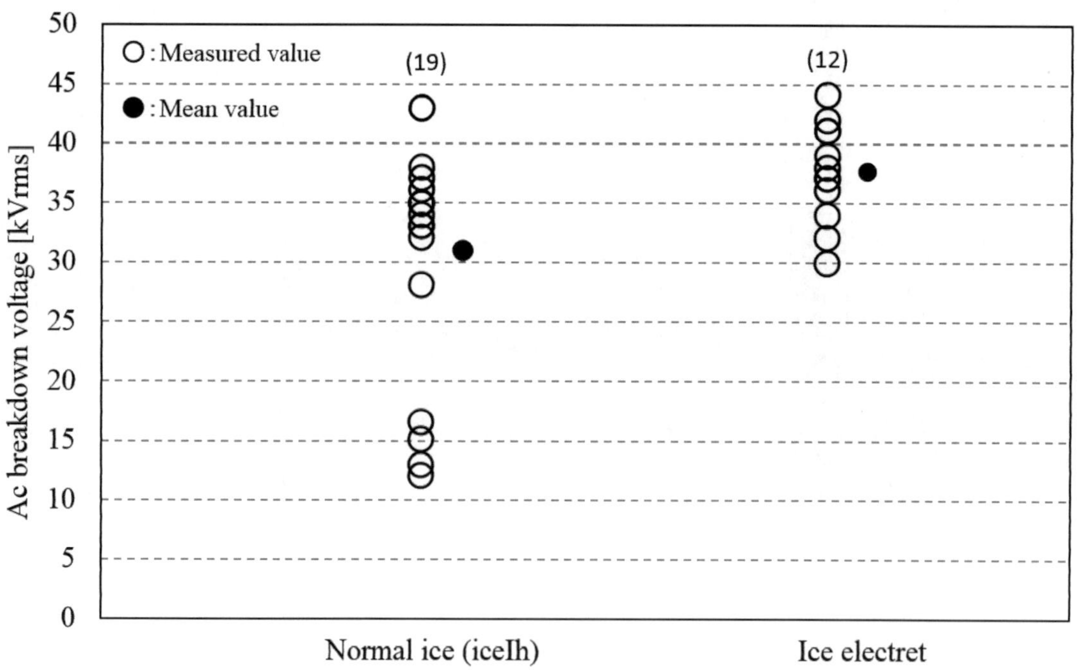

Figure 6. Breakdown voltage of normal ice and ice electret.

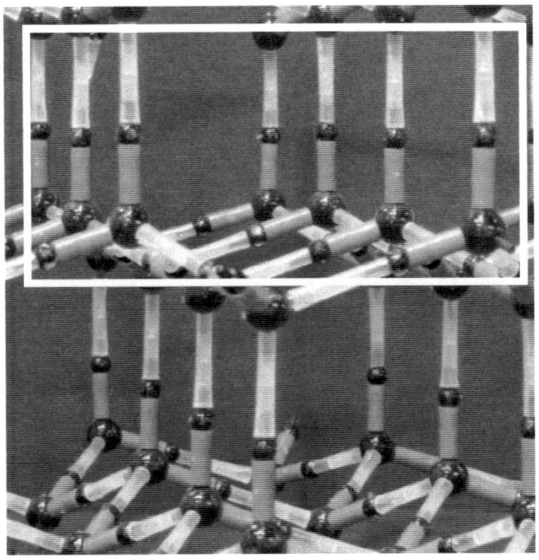

Figure 7. Crystal structure model of iceXI. Larger spheres are oxygen atoms, smaller spheres are hydrogen ions, dark poles indicate covalent binding and pale poles indicate hydrogen bonds.

Figure 8. Crystal structure model of iceIh. Spheres and poles models are same as figure 7.

model of the crystal structure of iceXI. The protons always occupy the upper positions to oxygen atoms. It is easy to understand the characteristics of proton arrangement as shown in the white frame of figure7. The structure of iceXI is known as the most stable structure of iceIh (normal ice). Figure 8 shows a model of the crystal structure of iceIh. There are no rules for proton distribution in iceIh. Protons exist at arbitrary locations. The crystal structure, including the proton distribution, in iceXI is regularly distributed compared with that of iceIh [6, 12]. Note the white frame inside of each figure. It easy to find the difference in proton arrangements between figure 7 and 8. We considered that the crystal structures of ice electret and iceXI are almost the same because the ice electret maintains a polarization similar to that of iceXI. Namely, the proton arrangement of normal ice and ice electret differ. Therefore, there is a possibility that the proton arrangement of ice influences the breakdown voltage.

4. Conclusions

We studied the effect of proton arrangement on the ac breakdown voltage of normal ice and ice electret at liquid nitrogen temperatures. The main results obtained are summarized as follows.

The maximum values of ice electret and normal ice were almost the same. By contrast, the minimum value of ice electret was 2.5 times larger than that of normal ice. For normal ice, a breakdown occurred even in the range of 12–17 kVrms, but the breakdown of ice electret did not occur below 30 kVrms. From these results, we considered that the electrically weak part of normal ice improved because changes in the proton arrangement of ice were made under a direct electric field application to normal ice.

References

[1] Shiji Y, Muramoto Y and Shimizu N, 2007, *IEEE T Dielect el in*, **14**, pp.296–301.
[2] Tsuchiya R, Muramoto Y and Shimizu N, 2013, Proc. Int. Conf. on Solid Dielectronics (Bologna) pp.533–536.
[3] Tsuchiya R, Muramoto Y and Shimizu N, 2016, *IEEJ Trans. FM*, **136**, pp.54–63 (in Japanese).

[4] Muto Y, Muramoto Y and Shimizu N, 2013, Proc. Int. Conf. on Solid Dielectronics (Bologna) pp.537–540.
[5] Oshika Y, Tsuchiya Y, Hashimoto T and Muramoto Y, 2017, "The depolarization current properties of ice electret at cryogenic temperature", TEION KOGAKU, **52**, in press (in Japanese).
[6] Fukazawa H, Hoshikawa A, Ishii Y, Chakoumokos BC and Fernandez-Baca JA, 2006, *Astrophys.J. Letters*, **652**, pp.L57–L60.
[7] Kawada S and Dohata H, 1985, *J.Phys.Soc.Jpn*, **54**, pp.477–9.
[8] Haida O, Matsuo T, Suga H and Seki S, 1972, "Relaxational Proton Ordering and Glassy Crystalline State in Hexagonal Ice", *Proc.Japan Acad. Ser B, Physical and Biological*, **48**, pp.489–94.
[9] Fujino K, 1967, "Electrical Properties of Sea Ice", *Physics of Snow and Ice*, **1**, pp.633–48 (in Japanese).
[10] Mathes K, 1965, "Electrical properties of cryogenic liquid and of ice at cryogenic temperature", IEEE 1965 Annual Report of Conf. on Electrical Insulation and Dielectric Phenomena, pp. 20–24.
[11] Gelin H and Stubbs R, 1965, *J. Chem. Phys.*, **42**, pp.967–71.
[12] Malemkov G, 2009, *J. Phys.-Condens. Mat*. **21**, 283101.

Irradiation effect of the insulating materials for fusion superconducting magnets at cryogenic temperature

Koji Kobayashi, Yoko Akiyama, Shigehiro Nishijima

Division of Sustainable Energy and Environmental Engineering, Graduate School of Engineering, Osaka University, 2-1 Yamadaoka, Suita, Osaka, 565-0871, Japan

E-mail: k-kobayashi@qb.see.eng.osaka-u.ac.jp

Abstract. In ITER, superconducting magnets should be used in such severe environment as high fluence of fast neutron, cryogenic temperature and large electromagnetic forces. Insulating material is one of the most sensitive component to radiation. So radiation resistance on mechanical properties at cryogenic temperature are required for insulating material. The purpose of this study is to evaluate irradiation effect of insulating material at cryogenic temperature by gamma-ray irradiation. Firstly, glass fiber reinforced plastic (GFRP) and hybrid composite were prepared. After irradiation at room temperature (RT) or liquid nitrogen temperature (LNT, 77 K), interlaminar shear strength (ILSS) and glass-transition temperature (Tg) measurement were conducted. It was shown that insulating materials irradiated at room temperature were much degraded than those at cryogenic temperature.

1. Introduction

In ITER, high temperature plasma is confined by strong magnetic field of superconducting magnets. The superconducting magnets are operated under such severe environment as high fluence of fast neutron (10^{22} n/m^2, E > 0.1 MeV), cryogenic temperature (liquid helium temperature, 4.2 K) and large electromagnetic forces, especially large shear strength up to 45 MPa [1] [2]. It has been reported that insulating materials are the most sensitive to the irradiation among the components of superconducting magnets. So, radiation resistance on mechanical properties at cryogenic temperature is required for insulation materials.

The insulating materials are hybrid composites which consist of glass cloth, polyimide film and organic polymer. Previous studies have investigated irradiation effect of hybrid composite and glass fiber reinforced plastic (GFRP) which consists of glass cloth and organic polymer. As a result, it was shown that the insulating material using compound of cyanate ester and epoxy resin have favorable radiation resistance on mechanical strength even after the irradiation simulated in ITER design [3] [4]. However, most of those studies have investigated the irradiation effect of insulating materials after room temperature (RT) irradiation. In ITER, actual irradiation environment of insulating materials is cryogenic temperature. In addition, it is considered that the temperature of the magnet components including the insulating materials change intermittently for magnets maintenance. The purpose in this study is to evaluate the irradiation effects of insulating materials at cryogenic temperature and those of temperature change after irradiation.

2. Experiment

2.1. Materials

To evaluate the mechanism of the irradiation induced degradation, we used the epoxy resin which is relatively sensitive against radiation compared with the compound of cyanate ester and epoxy, in this study. Epikote828 (Diglycidyl Ether of Bisphenol-A, DGEBA, Mitsubishi Chemical Corporation) as an epoxy resin and Baxxodur EC301 (Polyetheramin D230, Mitsui Fine Chemicals, Inc.) as a hardener were prepared. Figure 1 shows the chemical structures of Epikote828 and D230. GFRP and hybrid composite were fabricated by using the resins. Boron-free S-glass cloth was used in GFRP and hybrid composite.

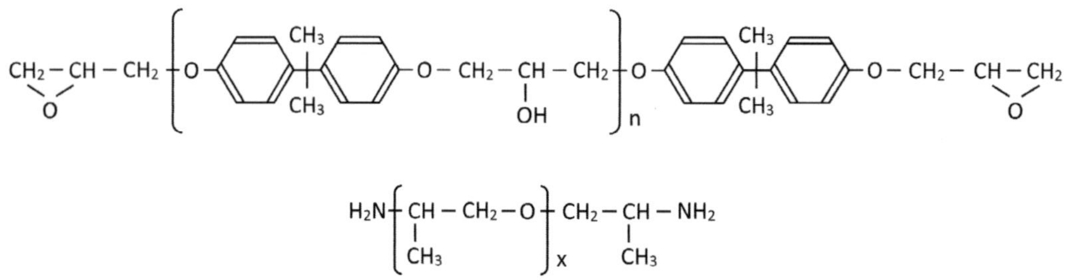

Figure 1. Chemical structure of DGEBA and D230

2.2. Sample preparation

Epikote828 and D230 were mixed at the epoxy equivalent (100 : 32.3). In the case of GFRP, 45 glass cloth sheets were laminated and impregnated with the resin under vacuum. In the case of hybrid composite, 29 glass cloth sheets and 28 polyimide film sheets were laminated. After vacuum impregnation, they were cured at 70 ℃ for 3 hours and then at 110 ℃ for 2 hours. Composites were cut into double-notch specimens for interlaminar shear strength (ILSS) test after curing. Figure 2 shows the shape of specimen for ILSS test. As the double-notch specimen's thickness was about 6 mm, there were 7 or 8 layers of the glass cloth per millimeter thickness in the case of GFRP, and about 4 or 5 layers in the case of hybrid composite. Volume fiber content (V_f) was 41.09 % in the case of GFRP, and 26.22 % in the case of hybrid composite as observed value, respectively. By similar method, epoxy plate which consists of only matrix resin was made for Dynamic Mechanical Analysis (DMA). The plate was cut into 50 mm × 10 mm × 1.5 mm in size.

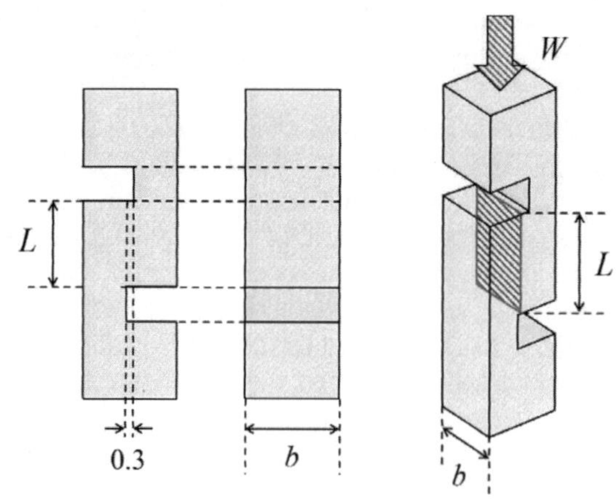

Figure 2. Shape of specimen for ILSS test

IOP Conf. Series: Journal of Physics: Conf. Series **897** (2017) 012007 doi:10.1088/1742-6596/897/1/012007

2.3. Irradiation and test procedures

Those specimens were irradiated by gamma-ray in ^{60}Co gamma-ray irradiation facility, Research Laboratory for Quantum Beam Science, Institute of Scientific and Industrial Research, Osaka University. Irradiation was conducted at RT under air atmosphere or at liquid nitrogen temperature (LNT, 77 K) in liquid nitrogen. Dose rate was 71.4 kGy/h and absorbed dose was 10 MGy. Figure 3 shows the double-notch specimens of GFRP and hybrid composite before and after irradiation.

ILSS test was performed by Shimadzu Autograph AG-X10kN (Shimadzu Corporation). Table 1 shows the irradiation temperature and thermal history of specimens. In condition A, the samples was irradiated at RT. After irradiation, they were cooled down to LNT and then ILSS was conducted at LNT. In condition B, the samples was irradiated at LNT. After irradiation, they were warmed up to RT and then cooled down to LNT again. ILSS test was conducted at LNT. In condition C, the samples was irradiated at LNT. After irradiation, they were kept at LNT and then ILSS test was conducted at the same temperature. ILSS was calculated by following formula;

$$\tau = \frac{W}{bL} \tag{1}$$

where, τ is interlaminar shear stress, W is load, b is depth of specimen and L is notch distance, respectively.

DMA was conducted by Advanced Rheometer AR1000 (TA Instruments). Dynamic storage modulus (G') and dynamic loss modulus (G'') were measured, and then the ratio G''/G' defined as $\tan\delta$ was calculated. The temperature where the $\tan\delta$ became maximum was defined as glass-transition temperature (Tg). The samples for DMA test irradiated at RT or LNT. The number of samples was 2 for each irradiation condition.

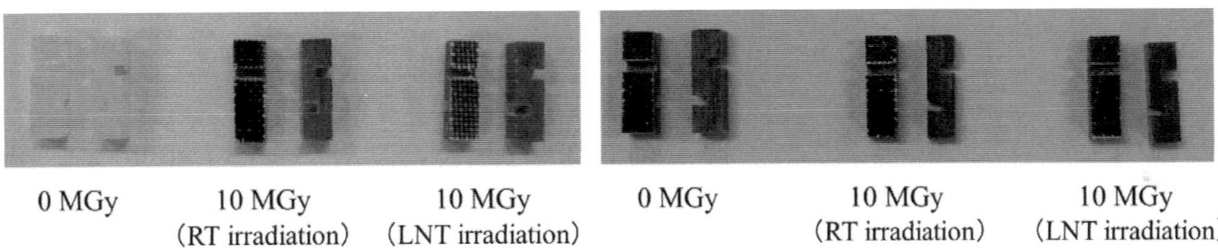

Figure 3. Specimens before and after irradiation (left : GFRP, right : Hybrid composite)

Table 1. Temperature condition of specimen in ILSS test

Condition	Temperature		
	Irradiation	Thermal History	ILSS test
A	RT	cooled down	LNT
B	LNT	warmed up to RT, and then cooled down again	LNT
C	LNT	constant	LNT

3. Results and discussion

3.1. ILSS test

Figure 4 shows the ILSS of GFRP and hybrid composite before and after irradiation. Concerning the GFRP, only the RT irradiated specimen shows the degradation. That is, the LNT irradiation does not degrade the ILSS of GFRP. The same phenomenon was observed in the hybrid composite specimens. This result indicated that the mechanical strength of the insulating material was not degraded by cryogenic irradiation whereas that was degraded by RT irradiation.

In the case of LNT irradiation, the significant difference was not induced between by raising temperature after irradiation (condition B) and by keeping at LNT (condition C). In addition, ILSS of hybrid composite specimens irradiated at RT shows larger degradation than those of GFRP. This result suggests that the interfacial strength between the resin and the polyimide film is more susceptible to radiation than those between the resin and the glass fiber. As a cause, it is considered that glass cloth has a larger adhesion area per unit area with the matrix resin compared with polyimide film, because the resin also flows into the gaps of the glass fibers by its permeability. Thereby the interface between the resin and the polyimide film had weaker strength, and then destruction occurred at the interface between the resin and the polyimide film.

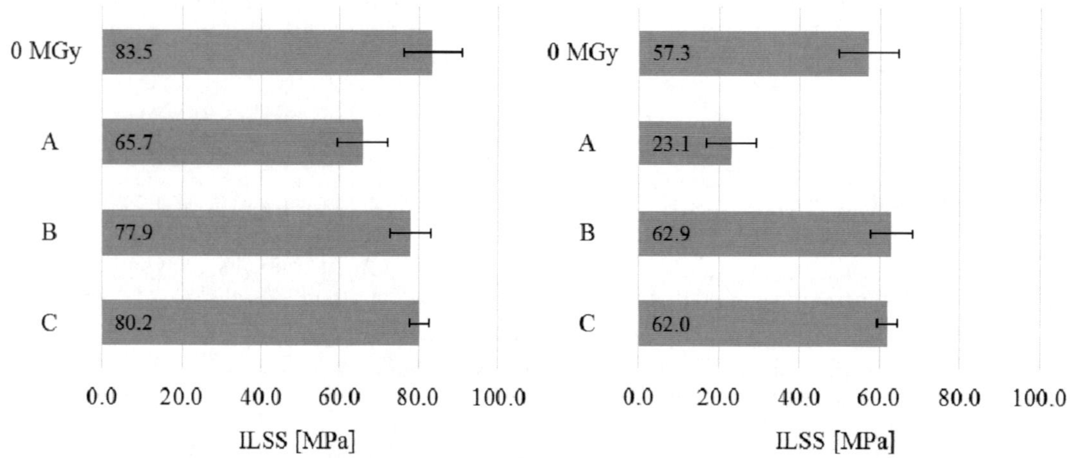

Figure 4. ILSS before and after irradiation (left : GFRP, right : Hybrid composite)

3.2. DMA

Figure 5 shows the result of Tg measurement before and after irradiation. The Tg was slightly decreased by the LNT irradiation. The RT irradiation induced the large degradation in Tg. Generally, Tg decreases when the molecular chain scission occurs. Therefore, it indicates that the scission occurs more frequently by RT irradiation compared with LNT irradiation. In other words, it was shown that the molecular chain scission was hardly induced by the cryogenic irradiation, in contrast with RT irradiation.

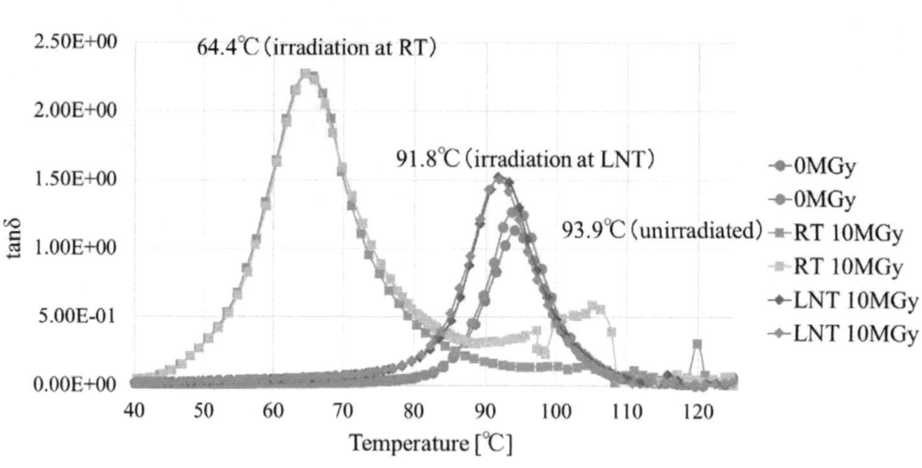

Figure 5. Tg before and after irradiation

The reason why the degradation of the insulating material does not occur markedly at LNT compared with that at RT is considered that the molecular chain motion affects the irradiation effect. Tg is the temperature where Micro-Brownian motion starts and then the molecular motion is frozen lower than Tg. At higher than Tg, G-value follow the Arrhenius equation, whereas lower than Tg G-value shows constant, because the hydrogen radicals generated by irradiation recombine without attacking other molecular chains. This phenomenon is called a "cage effect." Due to cage effect, it has been considered that G-value is constant lower than Tg [5]. However, as shown in Figure 5, insulating material is greatly degraded by RT irradiation although the Tg of the epoxy resin is higher than RT. Therefore, it is considered that the other molecular motion would affect the irradiation effect. In the epoxy resin, there is the temperature where the local molecular chain motion is taken place between RT and LNT [6]. That is, there is still local molecular chain motion at RT, though the molecular motion is frozen at LNT. In this study, insulating materials were degraded greatly by RT irradiation, but were hardly degraded by LNT irradiation. Therefore, it could be considered that the local molecular chain motion also affects the irradiation effect.

Another possible reason of degradation is oxygen effect. The irradiation at RT was performed in the air atmosphere. On the other hand, the irradiation at LNT was performed in liquid nitrogen where there was no oxygen. Oxygen is one of the factors affecting the irradiation effect. The experiments to evaluate the oxygen effect is now under planning.

In this study, DMA experiments were conducted only for matrix resin and we cannot discuss exactly if the local molecular motion affect the interface strength. But it is expected that local molecular motion will also affect the interfacial adhesion strength between resin and polyimide film. On the other hand, interface between the resin and glass fibres have strong bond by silane coupling agent that is independent from local molecular motion of the matrix resin. Above is considered as one of the reason why the irradiation effect on hybrid composite was larger than that on GFRP at RT. In future, we will validate the hypothesis by the experiment for irradiation effect on the interfacial bonding.

4. Conclusion

The purpose in this study is to evaluate the irradiation effect of insulating materials at cryogenic temperature especially the effect of temperature change after irradiation. Double-notch specimens made of GFRP or hybrid composite were irradiated at RT or LNT. The epoxy plate was prepared and also irradiated at RT or LNT. The ILSS test was made in GFRP and hybrid material, and Tg measurement was made in epoxy plate. By ILSS test, it was shown that ILSS was not degraded by LNT irradiation whereas ILSS was degraded by RT irradiation. In addition, it was also shown that ILSS was not changed by the temperature change between RT and LNT. Tg measurement showed that molecular chain scission in the matrix of insulating materials was not caused by LNT irradiation in contrast with RT irradiation. From these results, it was shown that insulating materials irradiated at cryogenic temperature was not degraded. As a cause of these results, it is considered that the local molecular motion could bring the deference. In order to verify the mechanism, the experiments eliminating the oxygen effects is now under planning.

Acknowledgments
This research was partly supported by NIFS Collaboration Research program (NIFS16KERA010).

References

[1] 2011 *Jounal Plasma and Fusion Research* **87** (Japan: Japan Society of Plasma Science and Nuclear Fusion Research)
[2] 2006 *ITER Design Description Document DDD 11 Magnet, ITER_D_22HV5L v2.2, Section 1. Engineering Description*
[3] Tomi Y, Mishima F, Akiyama Y, Izumi Y, Nishimura A and Nishijima S 2012 *IEEE Transactions on Appleid Superconductivity* **22** 7701104

[4] Li J, Wu Z, Huang C and Ki L 2014 *Fusion Engineering and Design* **89** pp 3112-3116
[5] Nishijima S and Okada T 1979 *Cryogenic engineering* **14** pp 99-109
[6] Ogata M, Kinjo O, Eguchi H and Kawata T 1989 *Kobunshi Ronbunshu* **46**(7) pp 413-419

Optimization of loading ratio of ErN as regenerator of 4K-GM cryocooler

T. Nakagawa[1], T. Miyauchi[1], T. Shiraishi[1], S. Seino[1], T. A. Yamamoto[1], Y. Fujimoto[2], S. Masuyama[3]

[1]Graduate School of Engineering, Osaka University, 2-1 Yamadaoka, Suita, Osaka 565-0871, Japan
[2]Institute of Laser Engineering, Osaka University, 2-6 Yamadaoka Suita, Osaka 565-0871, Japan
[3]Electoronic Mechanic Engineering Department, National Institute of Technology, Oshima College, 1091-1 Komatsu, Suo-oshima-cho, Oshima-gun, Yamaguchi 742-2193, Japan

nakagawa@mit.eng.osaka-u.ac.jp

Abstract. High purity erbium nitride (ErN) spheres with the size range of 150-180 μm and 180-212 μm were prepared by nitriding Er metal spheres with low oxygen content. The initial regenerator material of $HoCu_2$ on the cold end of the second regenerator column in 4K-GM cryocooler with nominal cooling power of 0.1 W at 4.2 K was replaced by ErN with different sizes. Higher cooling power was obtained when ErN of smaller size with lower oxygen content was used. We investigated the effect of partial replacement of $HoCu_2$ by ErN in the cold end side of second stage regenerator column on cooling power of 4K-GM cryocoolers. When ErN were substituted for 20 % of $HoCu_2$, the cooling power at 4.2 K reached 0.318 W. This value was 1.36 times as high as that of the cooling power of the GM cryocooler with commercially available regenerator arrangement. Therefore, use of ErN regenerator materials leads to the energy-saving and downsizing of 4K-GM cryocoolers.

1. Introduction

Gifford-McMahon (GM) cryocoolers are typical 4 K cryogenic refrigerators widely used for ultra high vacuum pumps, helium and hydrogen liquefaction, and for systems using superconducting magnets such as magnetic resonance imaging diagnosis devices and magnetic levitation trains. However, to obtain 1 W of cooling power at 4.2 K using GM cryocoolers, input power as much as several kW is needed. It is a very important problem to enhance their performances at cryogenic temperatures from the view of the energy saving. The 4K-GM cryocoolers have 2 stages of regenerators. By repeating compress and expansion of helium gas, cryogenic temperate is formed. The first-stage regenerator consists of stacked copper meshes, and the second consists of packed Pb and $HoCu_2$ spheres. The performance and efficiency of the 4K-GM cryocoolers depend on the second stage regenerator materials. The most important requirement for the regenerator materials are large specific heat around the working temperatures. To improve the efficiency, the researchers and engineers are seeking advanced regenerator materials [1-4].

We have reported rare earth nitrides (ErN, HoN, DyN, TbN and GdN) have large specific heat at cryogenic temperatures [5, 6]. Among them, ErN shows a peak of specific heat at 4 K. Thus, ErN was expected to be suitable materials for second stage regenerator of 4 K GM cryocoolers. In fact, we have prepared spherical ErN with different sizes and tried to use second stage regenerator for GM cryocooler with nominal cooling power of 0.1 W at 4.2 K [7]. We succeeded to form 4.2 K using the GM cryocooler with ErN regenerator. The cooling power of the GM cryocooler at 4.2 K became higher when the ErN spheres with smaller sizes were used. In previous works, we used Er spheres containing 1.10 wt. % of oxygen as starting materials to fabricate ErN spheres. The impurity level was too high. If higher purity ErN is used, the cooling power can be enhanced. In previous works, the volume ratio of Pb/ErN was fixed at 6/4. So, optimization of loading ratio of ErN as regenerator of 4K-GM cryocooler is needed for higher efficient work. Aim of this work is to find the most suitable arrangement of ErN with low oxygen content as regenerator materials for 4K-GM cryocoolers.

2. Experiment

Commercially available Er metallic spheres with 0.07 wt% of oxygen (Santoku Corporation) were classified by sieves. ErN spheres were synthesized by nitriding the Er metallic spheres with different sizes in the hot isostatic pressing method in nitrogen atmosphere (1550°C and 195 MPa of 99.9999% nitrogen). Formation of ErN was confirmed by the X-ray diffraction. The specific heat of ErN was measured by PPMS (Quantum Design). Cooling tests were performed using a commercially available two-stage GM cryocooler (Model SRDK-101D, Sumitomo Heavy Industries Ltd.) with nominal cooling power of 0.1 W at 4.2 K. The operating conditions were the standard ones provided by supplier. The second regenerator column consists of packed Pb spheres, 212-300 μm diameter, together with the cryogenic magnetic regenerator. The initial regenerator material of $HoCu_2$ (180-300 μm in diameter) was replaced by ErN spheres with different sizes. The weights of the ErN regenerator materials loaded in the column were measured. Oxygen content of Er metal, size range and loading ratio of the regenerator materials loaded in the second regenerator column are summarized in Table 1. By varying the input power to a heater attached to the cold head and monitoring the equilibrium temperature, *T*, cooling power (CP) versus *T* curves were obtained. The lowest achieving temperature was determined when CP = 0.

Table 1 Summary of regenerator materials in the second regenerator column of 4K-GM cryocooler

<table>
<tr><th>No.</th><th>Regenerator</th><th>Oxygen content of Er as starting material (wt%)</th><th>Size (μm)</th><th>Volume loading ratio Pb : $HoCu_2$: ErN</th></tr>
<tr><td>1#</td><td rowspan="3">ErN</td><td>1.10</td><td rowspan="2">180-212</td><td rowspan="3">6 : 0 : 4</td></tr>
<tr><td>2</td><td rowspan="2">0.07</td></tr>
<tr><td>3</td><td>150-180</td></tr>
<tr><td>4*</td><td>$HoCu_2$</td><td>--</td><td>180-300</td><td>6 : 4 : 0</td></tr>
<tr><td>5</td><td rowspan="3">ErN</td><td rowspan="3">0.07</td><td rowspan="3">150-180</td><td>8 : 0 : 2</td></tr>
<tr><td>6</td><td>7 : 0 : 3</td></tr>
<tr><td>7</td><td>5 : 0 : 5</td></tr>
<tr><td>8</td><td>$HoCu_2$</td><td>--</td><td>180-300</td><td>5 : 5 : 0</td></tr>
<tr><td>9</td><td rowspan="4">$HoCu_2$, ErN</td><td rowspan="4">--, 0.07</td><td rowspan="4">180-300, 150-180</td><td>5 : 1 : 4</td></tr>
<tr><td>10</td><td>5 : 2 : 3</td></tr>
<tr><td>11</td><td>5 : 3 : 2</td></tr>
<tr><td>12</td><td>5 : 4 : 1</td></tr>
</table>

#Ref. 7 * Commercially available condition

3. Results and discussion

In Fig. 1, the XRD pattern of the obtained nitride sample is shown. This graph indicates that monophase of ErN was obtained. In this figure, XRD pattern of ErN prepared from Er metal spheres with 1.10 wt% of oxygen, which was used in previous work, also appears. In the pattern of the ErN with high oxygen content, the peak assigned by erbium oxide is recognized around 28°.

Temperature dependence of specific heat of ErN with different oxygen content is given in Fig. 2.

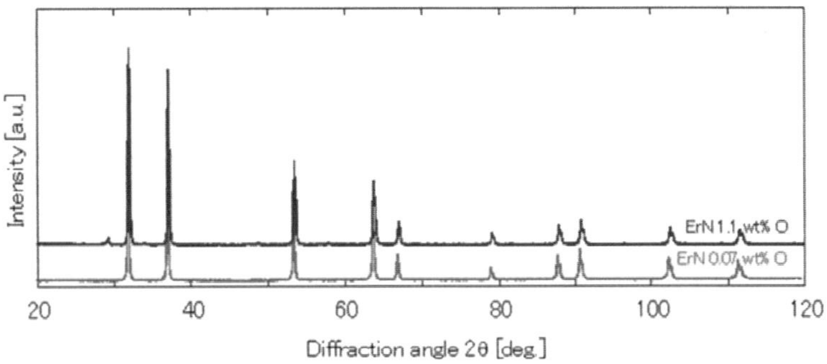

Fig. 1 XRD patterns of ErN with low and high oxygen contents.

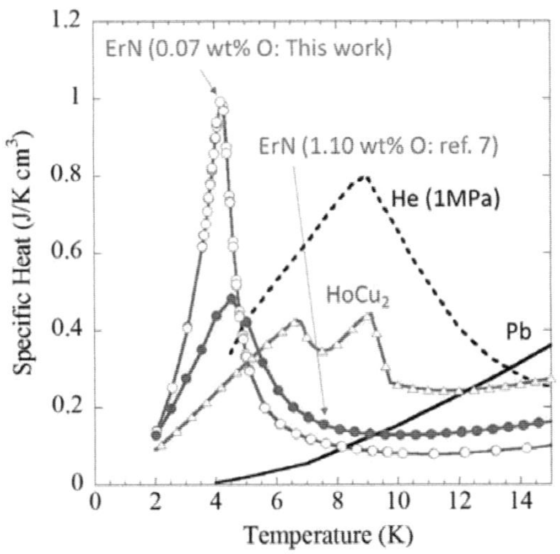

Fig. 2 Temperature dependence of heat capacities of the regenerator materials and He gas (1 MPa).

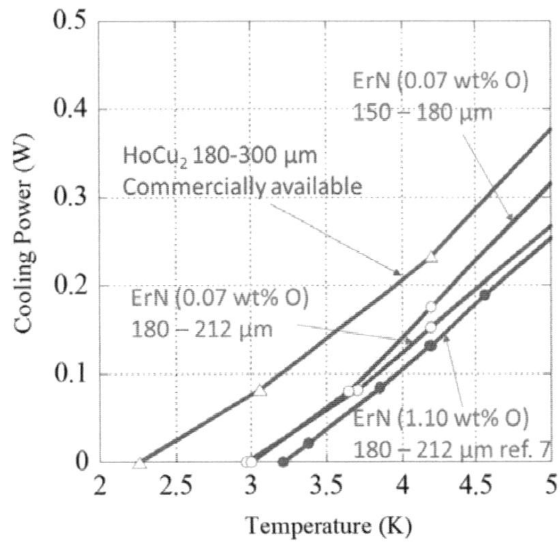

Fig. 3 Temperature dependence of cooling power of 4K-GM cryocooler using ErN with large oxygen content and low oxygen content with different sizes and $HoCu_2$.

Comparing to high oxygen content ErN, low oxygen content ErN shows steeper and larger peak of specific heat around 4 K. ErN specific heat of the low oxygen content indicates the essential specific heat of ErN. When the oxygen content is large, the oxygen atoms solute interstitially in nitride lattice. Thus, the peak of specific heat of ErN was broadened. The oxygen content is important factor to prepare ErN with large heat capacity.

Figure 3 indicates temperature dependence of CP of the GM cryocooler, in which ErN were loaded when volume ratio of Pb to ErN was 6 to 4. Higher CP was obtained at 4.2 K when ErN with lower oxygen content with the same sphere size was used. Compering to the difference of size, higher CP was obtained using smaller size of ErN. However, higher CP was obtained when $HoCu_2$ which were loaded in commercially available GM-cryocoolers was used.

To find the optimum volumetric loading ratio of ErN (0.07 wt. %, 150-180 μm) in second regenerator column, the ratio x (Pb : ErN = 10 - x : x) was changed 2 to 5. Figure 4 shows CP at 4.2 K as a function of loading ratio x of ErN. The CP at 4.2 K increased with increasing the loading ratio of ErN. However, the results were inferior compared with a case using $HoCu_2$. Figure 5 shows the loading ratio x dependence of the achieving lowest temperature. The lowest temperature is falling down with increasing loading ratio of ErN. However, the results were also inferior compared with a case using $HoCu_2$.

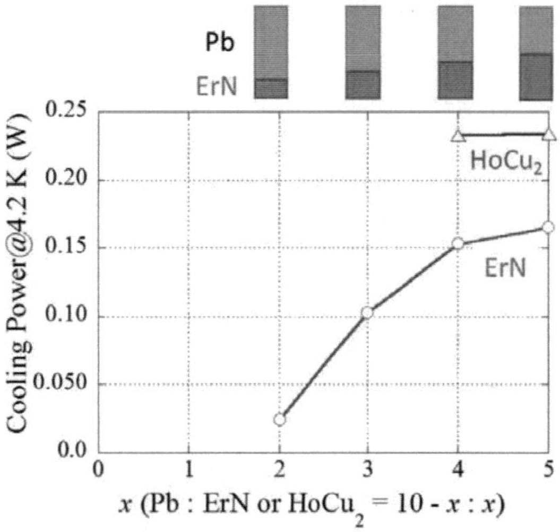

Fig. 4 Cooling power of 4K-GM cryocooler using combination of Pb and ErN or $HoCu_2$ with different loading ratio *x*.

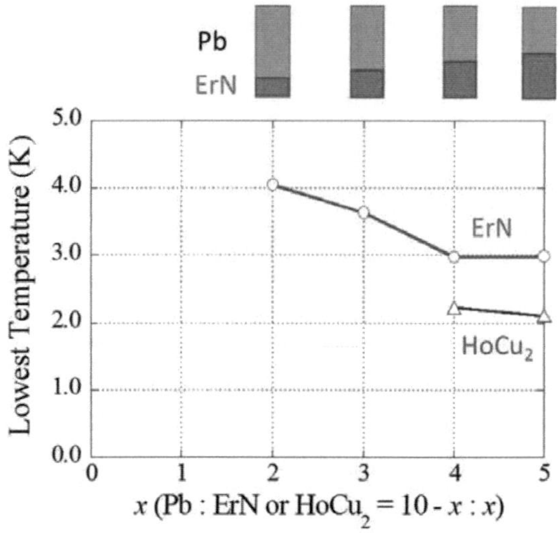

Fig. 5 Achieving lowest temperature of 4K-GM cryocooler using combination of Pb and ErN or $HoCu_2$ with different loading ratio *x*.

We explain the reason why inferior CP was obtained at 4.2 K using ErN in spite of higher specific heat of ErN at 4.2 K comparing to that of $HoCu_2$. In Fig. 2, heat capacities of $HoCu_2$, He gas (1 MPa), and Pb as a function of temperature are also given. He gas shows large heat capacity in the temperature range from 5 to 11 K and a peak around 8 K. $HoCu_2$ can reserve heat in this temperature though ErN and Pb cannot reserved enough. Thus, combination of only ErN and Pb does not realize higher CP than that of $HoCu_2$ and Pb.

To compensate small specific heat of ErN and Pb around the peak temperature of the specific heat of the He gas, $HoCu_2$ was inserted between ErN and Pb. Because both ErN and $HoCu_2$ show the highest CP at $x = 0.5$ (Pb: ErN or $HoCu_2 = 10 - x : x$) as shown in Fig. 4, the loading ratio of Pb was

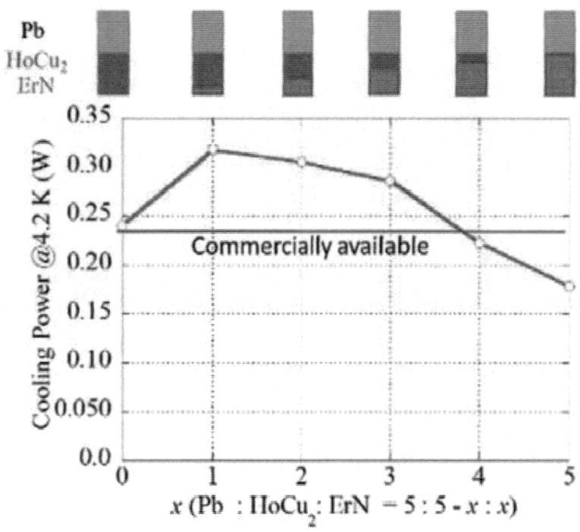

Fig. 6 Cooling power of 4K-GM cryocooler using combination of Pb, $HoCu_2$ and ErN with different loading ratio *x*.

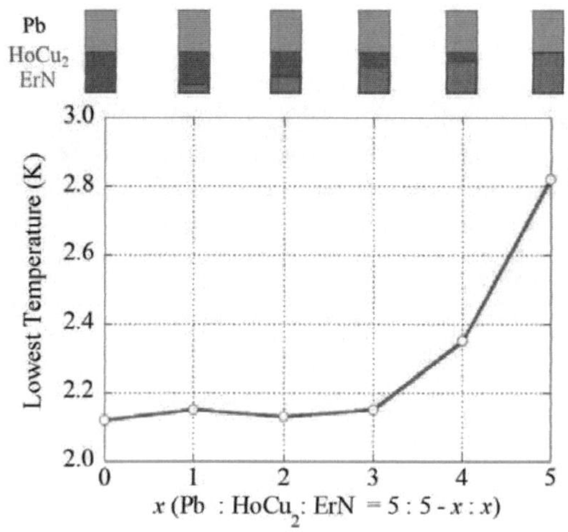

Fig. 7 Achieving lowest temperature of 4K-GM cryocooler using combination of Pb, $HoCu_2$ and ErN with different loading ratio *x*.

fixed at 5 and those of ErN and $HoCu_2$ were changed. Figure 6 shows the CP at 4.2 K as a function of loading ratio x (Pb : $HoCu_2$: ErN = 5 : 5 – x : x). The peak value can be found at $x = 1$ where the CP reaches 0.318 W. The loading ratio dependence of achieving lowest temperature is given in Fig. 7. The achieving lowest temperature is almost constant until loading ratio x becomes 3. When x exceeds 3, the lowest temperature rises.

The fact that only replacement of 20% of $HoCu_2$ by ErN leads to increasing rate on the CP of 36% against commercially available regenerator alignment allows the energy-saving and downsizing of 4K-GM cryocoolers.

4. Summary

To enhance cooling power at 4.2 K of GM cryocoolers, ErN spheres were prepared for regenerator materials.

- Large specific heat with a steep peak was observed for ErN spheres with low oxygen content.
- Higher cooling power was obtained using ErN spheres with lower oxygen content.
- When 20% of $HoCu_2$ were replaced by ErN at cold end in the second regenerator column, the cooling power at 4.2 K reaches 0.318 W. This value was 1.36 times as high as that of the GM cryocooler with commercially available regenerator arrangement (using $HoCu_2$).

References

[1] Sahashi M, Tokai Y, Kuriyama T, Nakagome H, Li R, Ogawa M and Hashimoto T 1990 *Adv. Cryogenic Eng.* **35** 1175
[2] Bischof J, Diviš M, Svoboda P and Smetana 1989 Z, *Phys. Status Solidi* **114** K229
[3] Qiu L M, Numazawa T and Thummes G 2001 *Cryogenics* **41** 693
[4] Numazawa T and Satoh T 2003 *Cryocooler* **12** 397
[5] Nakagawa T, Arakawa T, Sako K, Tomioka N, Yamamoto T A, Kusunose T, Niihara K, Kamiya K and Numazawa T 2006 *J. Alloys & Comp.* **408-412** 191
[6] Nishio S, Nakagawa T, Arakawa T, Tomioka N, Yamamoto T A, Kusunose T, Niihara K, Numazawa T and Kamiya K 2006 *J. Appl. Phys.* **99** 08K901
[7] Nakano T, Masuyama S, Hirayama Y, Izawa T, Nakagawa T, Fujimoto Y, Irie T, Nakamura E and Yamamoto T A 2012 *Appl. Phys. Lett.* **101** 251908

IOP Conf. Series: Journal of Physics: Conf. Series **897** (2017) 012009 doi:10.1088/1742-6596/897/1/012009

Low-temperature hydrogen absorption in metallic nanocontacts studied by point-contact spectroscopy measurements

H. Takata[1], M. S. Islam[1], K. Ienaga[1], Y. Inagaki[1], K. Hashizume[2] and T. Kawae[1]

[1] Department of Applied Quantum Physics, Kyushu University, Fukuoka 819-0395, Japan
[2] Department of Engineering Science, Kyushu University, Fukuoka 816-8580, Japan

Abstract. We report on hydrogen (H) and deuterium (D) atoms absorption below $T = 20$ K in metallic palladium (Pd) via quantum tunnelling (QT). When a small bias voltage is applied between Pd nanocontacts that are immersed in liquid H_2 (D_2), the differential conductance spectra measured by point-contact spectroscopy change enormously. The results indicate H (D) absorption in Pd nanocontacts at the temperature where H (D) absorption due to thermal hopping process is not expected, and can be explained by QT. The QT occurs when the energy level of the potential well trapping the H (D) atom coincides with those not trapping the H (D) atom, and is assisted by phonons induced by ballistic electrons.

1. Introduction

The use of fossil fuels causes devastating environmental impacts such as global warming due to carbon dioxide emissions. In addition, their supplies will deplete in near future. These situations accelerate the research on the engineering application of hydrogen (H), which is expected to be an ideal replacement for fossil fuels, because H_2 has high energy density by weight and its combustion products, H_2O, is environmentally clean. A large amount of H atoms are absorbed in metal hydride, therefore which is studied as carrier and storage for H. Although intensive efforts have been devoted to understand the absorption and diffusion process of H atom in hydride, it is unclear because of its strong quantum character arising from the lightest mass.

Quantum tunnelling (QT), which is one of the most fundamental concepts in quantum mechanics, is essential for H absorption and diffusion process in metal [1]. QT of neutral atoms, including H atom in metal, has been studied with a variety of experiments such as quasi-elastic neutron scattering and nuclear magnetic resonance [2,3]. However, QT is experimentally difficult to detect, in contrast to QT of electronic systems which can be detected directly via transport measurements. Hence, most studies on QT of H are discussed in terms of the temperature dependence of the reaction rate or the relaxation time [1,4,5]. In the thermal region, the temperature dependences of these processes obey the Arrhenius equation; while in the quantum region, there is no temperature dependence. These indicate that QT of H is ambiguous when the temperature dependences are complex.

We focus on point-contact spectroscopy (PCS) to investigate low-temperature properties of hydrides [6]. For the contact smaller than the mean free path of conduction electrons, the electrons accelerated by the bias voltage are moved in the contact without energy dissipation (ballistic electron). Hence, we can explore the properties of hydride through the inelastic scattering process between the electrons and H atoms using PCS [7-10]. Moreover, we found that the inelastic scattering induces low-temperature H

E-mail: t.kawae.122@m.kyushu-u.ac.jp

IOP Conf. Series: Journal of Physics: Conf. Series **897** (2017) 012009 doi:10.1088/1742-6596/897/1/012009

absorption in palladium (Pd) nanocontacts from liquid H_2 [6].

In this paper, we report the effect of high bias voltage for H absorption in Pd nanocontacts. The results indicate that the maximum concentration of H is almost ~ 0.6, which is attained by applying the bias voltage larger than 60 mV. Additionally, we perform deuterium (D) absorption experiments to study the generality of the low temperature absorption. Although the H and D spectra show a difference because of the mass difference, the absorption process can be explained by the same scheme.

2. Experimental

The present experiments are carried out with a home-made cryostat attaching a mechanically controllable break junction (MCBJ) apparatus in vacuum and low-temperature environment, which prevents surface contamination and oxidation. Pd polycrystalline wire (99.998%, 0.2mm diameter) on a phosphor-bronze substrate is stretched by bending the substrate with mechanical force, which is generated by a screw and a piezo element as shown in figure 1. The size of nanocontact is changed continuously from atomic scale to micrometres and is kept constant over several hours by avoiding thermal input to the sample space [11, 12]. In order to immerse the nanocontacts in liquid H_2, the MCBJ apparatus is mounted in an inner cell with flexible bellows inside the vacuum chamber [13]. Note that the differential conductance measurements to detect H absorption and diffusion are performed in liquid H_2, preventing the temperature increase due to Joule heating in the nanocontacts during the measurements.

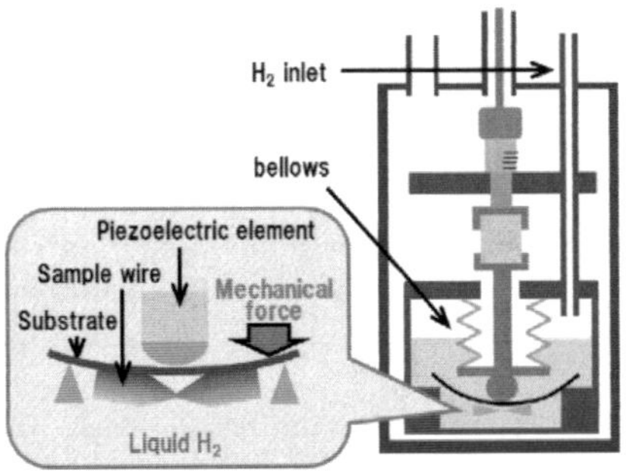

Figure 1. (Color online) Schematic illustration of MCBJ experimental cell. H absorption and diffusion experiments can be carried out in liquid H_2, avoiding the temperature increase due to Joule heating in the nanocontacts during the measurements.

The differential conductance dI/dV spectra are measured as a function of bias voltage using a lock-in technique (1 kHz modulation). d^2I/dV^2 spectra are obtained from the numerical derivative of dI/dV.

3. Results

Figures 2(a) and 2(b) show dI/dV spectra, and its numerical derivative of d^2I/dV^2 for pure Pd and $PdH_{0.6}$ nanocontacts, respectively. The spectra are measured in vacuum at $T = 18$ K, where the vertical axis is plotted in the universal conductance unit $G_0=2e^2/h$. Note that H atoms in $PdH_{0.6}$ are loaded at room temperature by the following procedure. Pd wire is exposed to 1 atm of H_2 gas in the inner cell at 300 K for several hours, making the H concentration to be $x \sim 0.6$. Then, the H_2 gas is evacuated from the inner cell at 80K to avoid H desorption from $PdH_{0.6}$ wire. Finally, we prepare the nanocontact by stretching $PdH_{0.6}$ wire at 18 K.

In pure Pd, a peak is observed at around the zero bias in dI/dV spectra, while a sharp peak due to electron-phonon scattering is observed at $|V| \sim 20$ mV in d^2I/dV^2 spectra. These features are consistent with other pure metals, indicating formation of clean nanocontacts without impurities on the surface. As shown in figures 2(a) and 2(b), the spectra are changed significantly from those of pure Pd by hydrogenation. The zero bias peak in dI/dV spectrum broadens. Moreover, in d^2I/dV^2 spectrum a new broad peak emerges at around $|V| \sim 60$ mV, while the peak at $|V| \sim 20$ mV is suppressed. From the neutron experiments, the first excited level of H atom trapped in fcc Pd octahedral site is estimated to be at $E \sim$ 60 meV from the ground state [14], which is consistent with the bias voltage of the new broad peak. This indicates that the excited energy level of H atom trapped in Pd can be measured with PCS.

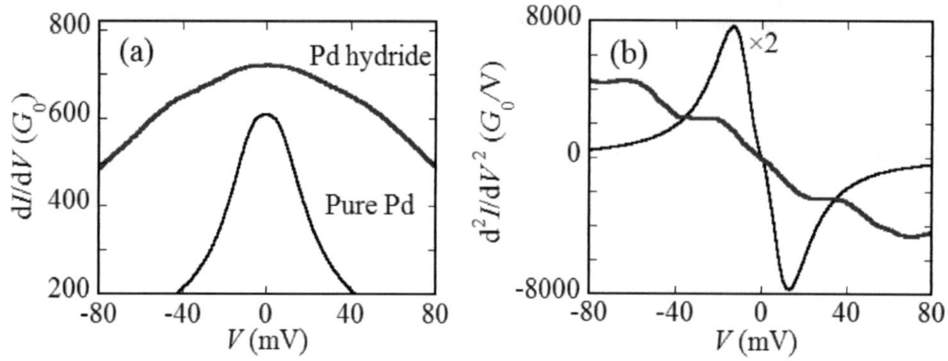

Figure 2. (Color online) (a) dI/dV spectra for pure Pd and $PdH_{0.6}$ nanocontacts in vacuum at 18 K, where the vertical axis is the universal conductance unit $G_0=2e^2/h$. (b) d^2I/dV^2 spectra, which are the numerical derivative of (a). The spectrum for pure nanocontact is scaled down one-half.

Simultaneously, the vibrational mode of H atom is induced by scattering with ballistic electrons accelerated by bias voltage. These results strongly suggest that H absorption and diffusion phenomena can be detected through PCS measurements.

Using PCS, we have performed H absorption and diffusion experiments into Pd nanocontacts from liquid H_2. After the pure Pd nanocontact is prepared at $T \sim 18$ K in vacuum, liquid H_2 is introduced in the inner cell. Then we start the differential conductance measurements with increasing the amplitude of the bias voltage from $|V| = 20$ mV to 100 mV with 20 mV step. After starting the measurements, the spectra shape changes gradually and finally becomes almost constant several hours later for all the bias amplitudes. Figures 3(a) and 3(b) show the bias voltage dependence of the dI/dV and d^2I/dV^2 spectra after finishing their changes. It is seen that the width of dI/dV curve broadens as the bias amplitude is increased. In addition, the peak value at $|V| \sim 20$mV in d^2I/dV^2 curve is rapidly decreased. Significantly, the shapes of the dI/dV and d^2I/dV^2 spectra for the bias voltage larger than 60mV are nearly the same with those of the spectra plotted in figures 2(a) and 2(b). These results demonstrate that H absorption into the nanocontacts is gradually progressed by applying the bias voltage although the thermal hopping of H atom is not expected. Furthermore, H concentration in the contact is almost saturated for the bias voltage larger than 60mV. From comparison with figures 2(a) and 2(b), the final H concentration is considered to be $x \sim 0.6$.

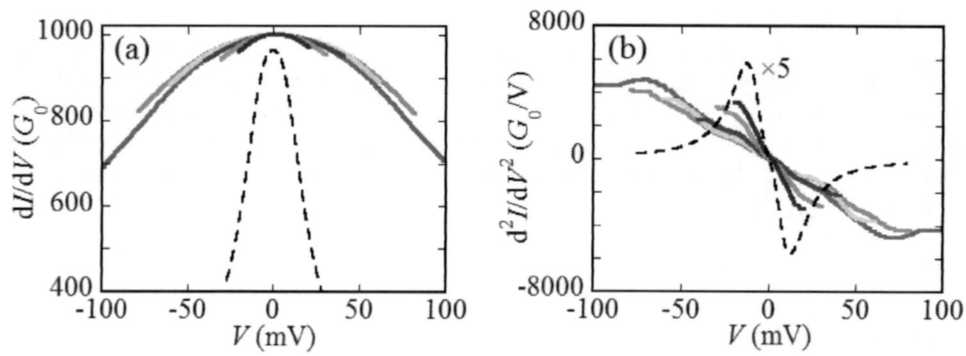

Figure 3. (Color online) The bias voltage dependence of dI/dV (a) and d^2I/dV^2 (b) spectra of Pd immersed in liquid H_2. The spectra in (a) and (b) are measured for the bias amplitude at $|V| = 20$, 40, 60, 80 and 100 mV, after finishing the spectra changes by H absorption into the nanocontacts. For comparison, the spectra for pure nanocontact (dashed line) are plotted in the figures. The d^2I/dV^2 curve is scaled down one-fifth.

Next, we perform D absorption experiments into Pd nanocontacts from liquid D_2 at T ~ 20 K to confirm the generality of H absorption observed in figures 3(a) and 3(b). The experimental procedure is the same with that of H absorption experiments. Figures 4(a) and 4(b) show the spectra change by D absorption, where the differential conductance is measured in liquid D_2 with the bias amplitude for $|V| \leqq 80$ mV. As in the case of H absorption measurements, the spectra shape changes gradually after starting the measurements and finally becomes almost constant several hours later, indicating that D absorption process is the same as H one. In figure 4(b), the new peak appears at $|V|$ ~ 40 mV, which is smaller than that at $|V|$ ~ 60 mV due to the H excitation. This shift is explained by the mass difference between H and D. Indeed, the value of V ~ 40 mV is in an excellent agreement with the first excited level of D trapped in fcc Pd octahedral site detected by the neutron experiments [14].

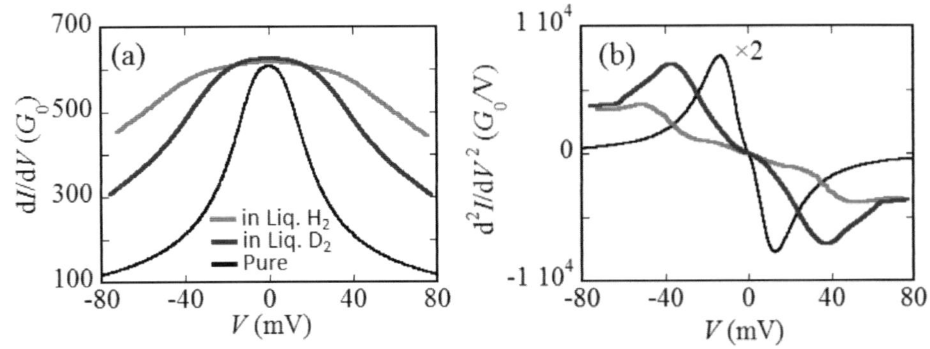

Figure 4. (Color online) (a) dI/dV spectrum for pure Pd nanocontact (black line), Pd nanocontacts immersed in liquid H_2 (red line) and liquid D_2 (blue line), where the bias voltage of ~ 80 mV is applied in liquid H_2 and D_2. (b) d^2I/dV^2 spectra for pure Pd nanoconatct, Pd nanocontacts immersed in liquid H_2 and liquid D_2. The spectrum for pure nanocontact is scaled down one-half.

Finally, let us discuss the origin of H/D absorption and diffusion below $T = 20$ K. Thermal hopping and QT are the two possible modes for the absorption and diffusion process. In thermal hopping mode, H atom trapped in an octahedral site of fcc Pd moves to a vacant site by jumping over the activation barrier E_{diff} ~230 meV for the absorption and diffusion process [15]. The diffusion coefficient is expressed by $D = D_0 \exp(-E_{diff}/k_BT)$ so that the thermal hopping is rapidly suppressed with decreasing temperature. Here, k_B is Boltzmann's constant. When $D_0 = 2.9 \times 10^{-7}$ m²·s for Pd [1], D is estimated to be ~10^{-70} m²·s at $T = 18$ K, implying that the thermal hopping process is prohibited in the present experiments. From these facts, it is reasonable to consider that QT is responsible for H/D absorption and diffusion below $T = 20$ K. We propose the following scenario to understand the results. When H/D atoms

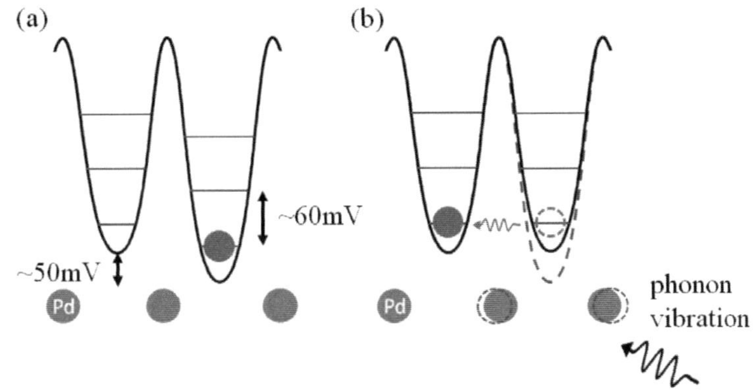

Figure 5. (Color online) (a) The self-trapped state of an H atom, where the energy level of the atom is ~ 50 meV lower than that of the vacant site. The first excited state is at ~ 60 mV from the ground state. (b) The diffusion process from the self-trapped state, where the H atom is excited by phonon vibrations generated by scattered inelastic electrons, and can tunnel to the neighbouring site.

are absorbed in metals, the lattice structure should be distorted, forming a "self-trap state" in Pd. In the case of H atoms, the energy level of the H trapped site is ~ 50 meV lower than the corresponding level in the H vacant one as shown in figure 5(a) [16]. Diffusion due to QT occurs when the energy level of H atom in the self-trapped state is equal to that of the vacant one as shown in figure 5(b). This process is induced by phonon vibrations, which are excited by the scattering with the ballistic electrons.

The present results indicate that when a small bias voltage is applied between the metallic surface and liquid H_2, H atoms are absorbed and diffused in metal due to QT. This would result in progress of H embrittlement even below 20 K, hindering safe utilization of liquid H_2. Obviously, further investigations on QT, e.g., material dependence, time dependence, are necessary to advance the industrial applications of liquid H_2.

4. Conclusion

We report on H and D atoms absorption in metallic Pd via QT. The differential conductance dI/dV and d^2I/dV^2 spectra show significant change when H or D atoms exist in Pd nanocontacts. Using the phenomena, we demonstrate that H and D atoms are absorbed in Pd nanocontacts from liquid H_2 (D_2) even below T = 20 K, when a small bias voltage is applied between Pd nanocontacts. The thermal hopping process is prohibited at this temperature, indicating that QT plays a crucial role for the absorption and diffusion of H and D. We explain that the QT occurs when the energy level of the potential well trapping the H atom coincides with those not trapping the H atom, and is assisted by phonons induced by ballistic electrons.

Acknowledgments

We would like to express our decent gratitude to the Japan Society for the Promotion of Science for providing the financial support by a Grant-in-Aid for Scientific Research (Grant Nos. 25220605, 25287076 and 26600102).

References

[1] Fukai Y 2005 *The Metal-Hydrogen System* , 2nd Edition (Springer, Berlin)
[2] Nelin G and Scöld K 1975 *J. Phys. Chem. Solids* **36** 117
[3] Arons R R, Bohn H G and Lütgemeier H 1974 *Solid State Commun.* **14** 1203 (1974).
[4] Minissale M, Congiu E, Baouche S, Chaabouni H, Moudens A, Dulieu F, Accolla M, Cazaux S, Manico G and Pirronello V 2013 *Phys. Rev. Lett.* **111** 053201
[5] Shannon R J, Blitz M A, Goddard A and Heard D E 2013 *Nature Chem.* **5** 745
[6] Ienaga K, Takata H, Onishi Y, Inagaki Y, Tsujii H, Kimura T and Kawae T 2015 Appl. Phys. Lett. **106** 021605
[7] Yanson I K and Shalov Yu N 1976 *Sov. Phys. JETP* **44** 148
[8] Jansen A G M, Mueller F M and Wyder P 1977 *Phys. Rev. B* **16** 1325
[9] Naidyuk Yu G and Yanson I K 2004 *Point-Contact Spectroscopy* (Springer, Berlin)
[10] Csonka Sz, Halbritter A, Mihály G, Shklyarevskii O I., Speller S and van Kempen H 2004 *Phys. Rev. Lett.* **93** 016802
[11] Ienaga, K, Nakashima N, Inagaki Y, Tsujii H, Honda S, Kimura T and Kawae T 2012 *Phys. Rev.* B **86** 064404 (2012).
[12] Ienaga K, Nakashima N, Inagaki Y, Tsujii H, Kimura T and Kawae T 2012 *Appl. Phys. Lett.* **101** 123114
[13] Ienaga K, Yokota T, Nakashima N, Inagaki Y, Tsujii H and Kawae T 2012 *J. Phys. : Conf. Ser.* **400** 042019
[14] Rowe J M, Rush J J, Smith H G, Mostoller M and Flotow H E 1974 *Phys. Rev. Lett.* **33** 1297
[15] Fukai Y and Sugimoto H 1985 *Adv. Phys.* **34** 263
[16] Flynn CP and Stoneham A M 1970 *Phys. Rev. B* **1** 3966

Thermal and magnetic properties of regenerator material Gd_2O_2S

K Matsumoto[1], T Numazawa[2], Y Ura[1], T Ujiyama[1] and S Abe[1]

[1] Department of Physics, Kanazawa University, Kanazawa 921-8111, Japan
[2] National Institute of Materials Science, Tsukuba 302-0003, Japan

k.matsu@staff.kanazawa-u.ac.jp

Abstract. Magnetic materials play a significant role in improvement of regenerative cryocooler performance, because they have high volumetric specific heat at magnetic transition temperatures. Gadolinium oxysulfide (Gd_2O_2S, GOS) that has an antiferromagnetic transition at 5 K improved the cooling performance of cryocoolers when it was used in colder side of the second stage regenerator operating below 10 K. Small magnetic susceptibility and specific heat insensitive to magnetic field is important in order to reduce influence of magnetic field on the performance of cryocooler. We measured magnetization and specific heat of ceramic GOS in magnetic field up to 5 T. The magnetization of GOS represented typical temperature dependence for antiferromagnetic materials and no metamagnetic transition was observed. As for specific heat of GOS, peak temperature decreased from 5.5 to 5.0 K with increasing magnetic field from 0 to 5 T and the transitions remained sharp in magnetic fields. Thermal conductivity of GOS was observed to have very small magnetic field dependence.

1. Introduction

Cryocoolers with lowest temperature of 4 K realized superconducting magnet systems and ^{3}He-^{4}He dilution refrigerators without liquid helium. In order to improve regenerative cryocooler performance, magnetic materials play a significant role in regenerator efficiency because they have high volumetric specific heat at magnetic transition temperature. Various magnetic materials such as Er_3Ni, $HoCu_2$ and $GdAlO_3$ were studied. Numazawa et al. developed ceramic gadolinium oxysulfide (Gd_2O_2S, GOS) that has an antiferromagnetic transition at 5 K in 2000's [1,2]. GOS has very high specific heat and sharp peak in comparison with Er_3Ni and $HoCu_2$ below 5 K. Ceramic GOS has high Vickers hardness. The cooling performance was much improved when GOS was used in colder side of the second stage regenerator that operated below 10 K.

Cryocoolers with magnetic regenerator materials are often used with superconducting magnet. Then, small magnetic susceptibility and specific heat insensitive to magnetic field is important in order to reduce influence of magnetic field. We measured magnetization and specific heat of ceramic GOS in magnetic field from 0 to 5 T. The magnetization of GOS represented typical temperature dependence for antiferromagnetic materials and no spin-flip transition was observed up to 5 T. As for specific heat of GOS, the peak temperature decreased from 5.5 to 5.0 K with increasing magnetic field from 0 to 5 T and the transitions remained sharp in magnetic fields. Thermal conductivity of GOS was measured in magnetic field 0 and 5 T. It was observed that the field dependence was very small. From these results, ceramic GOS was shown to be excellent magnetic regenerator material.

2. Experiments

A plate of ceramic gadolinium oxysulfide (Gd_2O_2S, GOS) was prepared and samples for various measurements are cut from this plates. Size and weights of samples are as follows: a plate of 3.89 mg for specific heat, a particle of 6.60 mg for magnetization, a rod with 5.29×3.03×21.8 mm^3 for thermal conductivity, and rectangular parallelepiped of 4.81×3.03×21.84 mm^3 for thermal expansion and magnetostriction. Figure 1 shows the photograph of the GOS samples used for thermal conductivity and dilation measurement.

Specific heat was measured with Quantum Design PPMS in magnetic fields of 0, 1, 3, and 5 T. Magnetization curves and temperature dependence were measured with Quantum Design MPMS. The measurements of the thermal conductivity were performed by a steady-state method with a homemade apparatus in 0 and 5 T. The temperature gradient was measured between two points on the rod sample when defined amount of heat flux was passing through the sample. The thermal conductivity was calculated with the temperature gradient, heat flux, and cross section of the sample rod in terms of Fourier's law. Thermal expansion and magnetostriction were measured using a homemade capacitive dilatometer that was installed in Quantum Design PPMS [3].

Figure 1. Photograph of GOS sample for thermal conductivity measurement (left) and magnetostriction (right).

3. Results and Discussion

3.1. Specific heat

Figure 2 shows the specific heat of ceramic GOS in 0, 3, and 5 T. Specific heat has a sharp peak at the antiferromagnetic transition temperature. The peak temperature decreased 0.2, 0.3 and 0.5 K with increasing field of 1, 3 and 5 T. This behaviour is typical for antiferromagnetic materials. The peak height was suppressed about 10% by 3 T magnetic fields but the sharpness is not varied significantly. The small transition temperature shift by magnetic field probably causes little effect on cryocooler performance, considering usual leak field of superconducting magnet. Gadolinium aluminium perovskite ($GdAlO_3$, GAP) is an antiferromagnetic material with the same rare-earth element and similar transition temperature of 3.8 K. Peak temperature of specific heat for GAP was reduced about 0.8 K by 2 T and suppression of peak height for GAP was much larger than that for GOS. In addition, peak of specific heat in GAP showed a significant broadening above 4 T. GOS has superior specific heat properties to GAP as regenerator material from the viewpoint of magnetic field effect.

The entropy of GOS was calculated from the specific heat using thermodynamic relation, $S(T) = \int_0^T \frac{C(T)}{T} dT$. Figure 3 represents the entropy of GOS in 0 T. As shown in Fig. 3, the entropy caused by magnetic spin system increases around the magnetic transition temperature. The entropy per one mole of Gd reaches about R$ln(8)$, where R is gas constant. The entropy shows saturation about 15 K. This represents that almost all the spin degree of freedom for Gd (J=7/2) is released at the antiferromagnetic transition and specific heat originated from lattice is small in this temperature region.

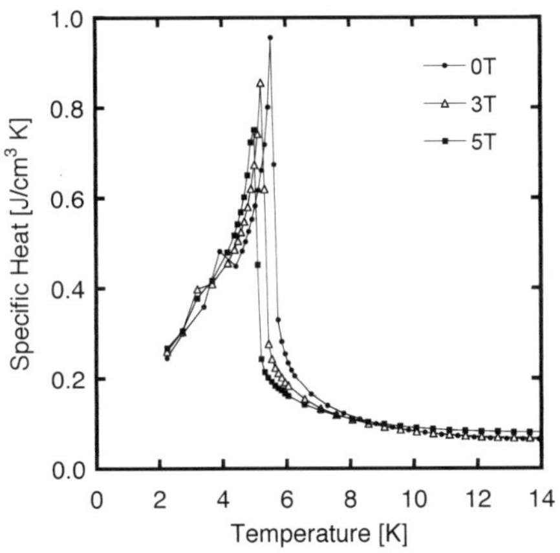

Figure 2. Specific heat of ceramic GOS in 0, 3, and 5 T.

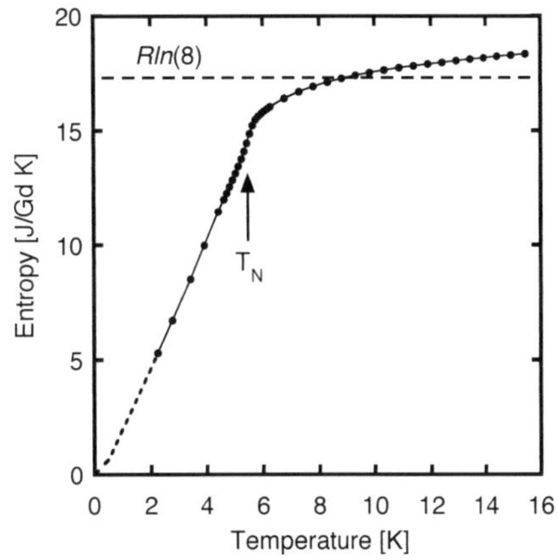

Figure 3. Entropy of ceramic GOS per one mole of Gd in 0 T.

3.2. Magnetization

Magnetization of GOS was represented as functions of temperature in 0.1, 3, and 5 T in Fig. 4. Magnetization is small and has a small peak the transition temperature as expected for an antiferromagnetic material.

Magnetization curve of ceramic GOS at 5 K was shown in Fig. 5. Those of $HoCu_2$ and ErNi at 5K [4] are plotted for comparison. Magnetization curves of GOS were measured in antiferromagnetic phase (4 K), around transition temperature (5 K), and in paramagnetic phase (6 K). There is no significant difference between them. Metamagnetic transition due to spin-flip of Gd ion wasn't observed up to 5 T in antiferromagnetic phase. Ferromagnetic ErNi has much larger magnetization than GOS. $HoCu_2$ shows spin-flip behaviour about 0.5 T. Comparing with volumetric magnetization between $HoCu_2$ and GOS, magnetization of GOS was about one third of $HoCu_2$ in 1 T.

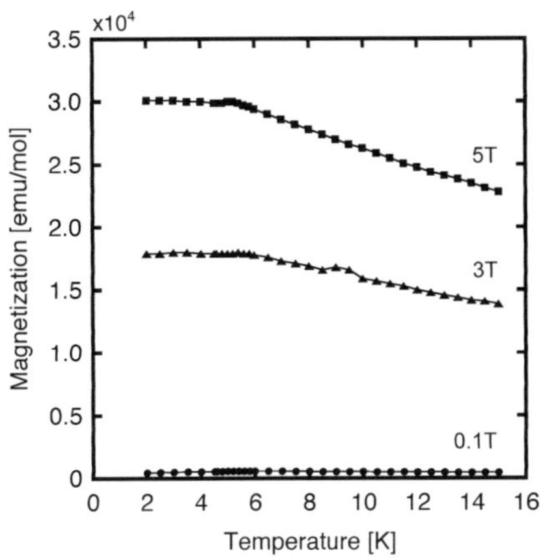

Figure 4. Magnetization of ceramic GOS as functions of temperature in 0.1, 3, and 5 T.

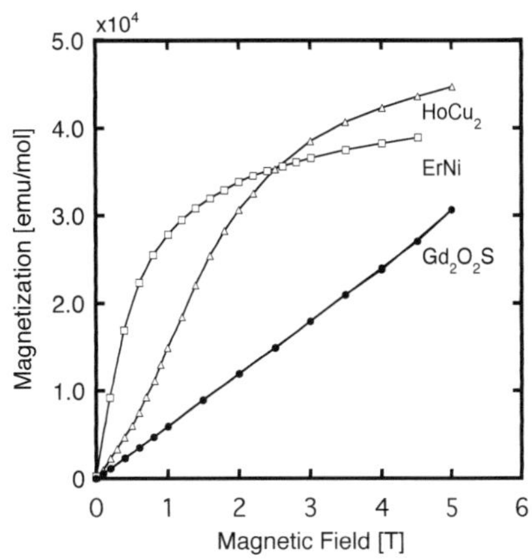

Figure 5. Magnetization curve of ceramic GOS at 5 K. Those of $HoCu_2$ and ErNi at 5K [4] are plotted for comparison.

3.3. Thermal conductivity

Figure 6 shows thermal conductivities of ceramic GOS in 0 T and some rare-earth magnetic intermetallic compounds considered as regenerator material for comparison. Thermal conductivity of GOS is comparable to those of intermetallic compounds [5, 6]. Temperature dependence of thermal conductivity of GOS was larger than those of intermetallic compounds. Thermal energy carrier difference between ceramics and intermetallic compounds may cause difference in temperature dependence.

Thermal conductivity of GOS in 0 T is close to that of dysprosium gallium garnet ($Dy_3Ga_5O_{12}$, DGG) [6]. However, DGG has large magnetic field dependence in thermal conductivity. Thermal conductivity in DGG was decreased about one order of magnitude by 5 T field. We measured thermal conductivities of GOS in 0 and 5 T and no change was observed. GOS has an advantage in thermal conductivity in magnetic field.

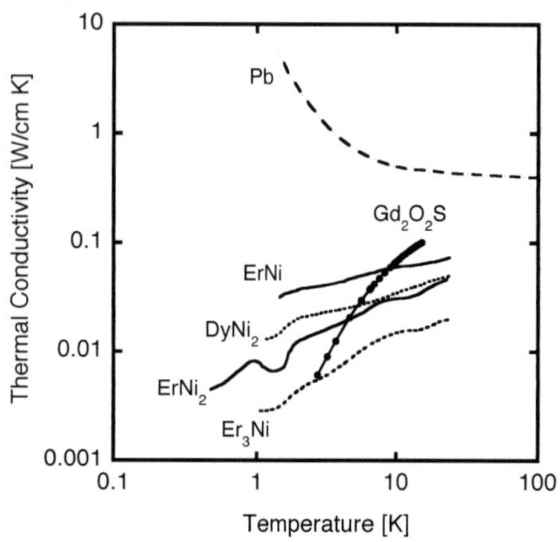

Figure 6. Thermal conductivity of ceramic GOS. Those of some rare-earth magnetic intermetallic compounds (ErNi, $DyNi_2$, $ErNi_2$, Er_3Ni) and lead [5] are plotted for comparison.

3.4. Thermal expansion and magnetostriction

Thermal contraction ($\Delta L/L$) of GOS was ~0.15% from room temperature to cryogenic temperature, that is about half of metal such as stainless steel and copper. Figure 7 shows thermal expansion of ceramic GOS below 15 K. Thermal expansion of ceramic GOS represents a slope change at the transition temperature. Additional contraction due to magnetic transition was $\sim 10^{-4}$% and much smaller than that of $ErAl_2$ that is metallic magnetocaloric material with ferromagnetic transition at 12 K [5].

Figure 8 shows magnetostriction of GOS in antiferromagnetic phase (2 and 4 K), around transition temperature (5.5 K), and in paramagnetic phase (6 K). Magnetstriction is plotted as $\frac{\Delta L}{L_{(0T)}} = \frac{L_{(B)} - L_{(0T)}}{L_{(0T)}}$. In all temperatures, GOS was expanded by magnetic field. In paramagnetic phase, magnetstriction is in proportion to squared field. In ordered phase, GOS expanded abruptly from 0 to 0.7 T. Above 0.7 T, magnetostriction showed almost same field dependence in paramagnetic phase. Magnetic domain motion is a possible reason of this large magnetostriction in low fields. $\Delta L/L_{(0T)}$ $\sim 2 \times 10^{-3}$% by 5 T is small. We observed no damage in GOS sample after applying magnetic field several times so that magnetostriction may not cause any trouble as regenerator material.

4. Summary

We measured specific heat, thermal conductivity, thermal expansion, and magnetostriction of ceramic GOS. As for specific heat, peak temperature decreased from 5.5 to 5.0 K with increasing magnetic field from 0 to 5 T and the transitions remained sharp in magnetic fields. Thermal conductivity of GOS was enough large to be used as regenerator material and had no magnetic field dependence. Magnetostriction was shown to be small and the ceramic GOS had no damage after cyclic field application. In conclusion, ceramic GOS has excellent magnetic and thermal properties as regenerator material.

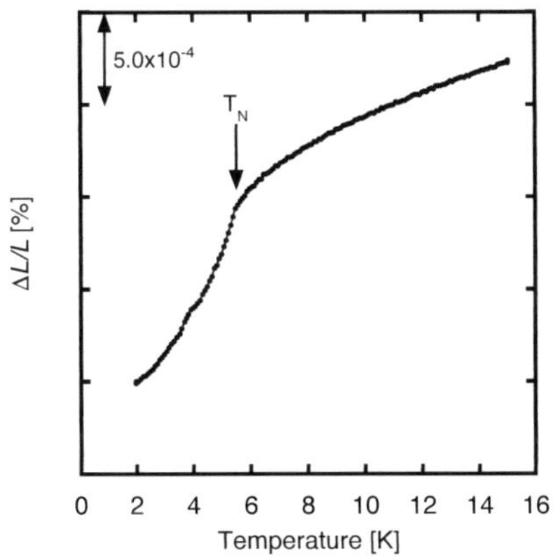

Figure 7. Thermal expansion of ceramic GOS around the transition temperature.

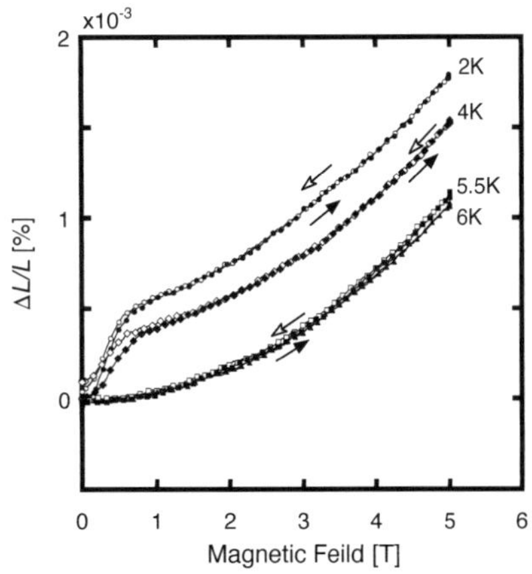

Figure 8. Magnetostriction of ceramic GOS in ordered phase (2 and 4 K), around the transition temperature (5.5 K) and para magnetic phase (6 K).

5. Acknowledgments

This work is supported by JSPS KAKENHI Grant Number 26420720, 26420890, and the Fujikura Foundation.

References

[1] Satoh T and Numazawa T 2003 *Cryocooler* **12** 397
[2] Numazawa T, Yanagitani T, Nozawa H, Ikeya Y, Li R and Satoh T 2003 *Cryocooler* **12** 473
[3] Iwakami O, Kawata N, Takeshita M, Yao Y, Abe S and Matsumoto K 2014 *Journal of Physics : Conference Series* **568** 032002
[4] Wang D, Li Y, Long Y, Ye R, Chang Y and Wan F 2007 *Journal of Magnetism and Magnetic Materials* **311** 697-701
[5] Ogawa M, Li R and Hashimoto T 1991 *Cryogenics* **31** 405
[6] Wikus P, Canavan E, Heine S T, Matsumoto K and Numazawa T 2014 *Cryogenics* **62** 150
[7] Kamiya K, Numazawa T, Koen T, Okano T, Matsumoto K and Nimori S 2004 *IEEE Transactions on Applied Superconductivity* **14** 1734

IOP Conf. Series: Journal of Physics: Conf. Series **897** (2017) 012011 doi:10.1088/1742-6596/897/1/012011

Magnetocaloric effect, thermal conductivity, and magnetostriction of epoxy-bonded $La(Fe_{0.88}Si_{0.12})_{13}$ hydrides

K Matsumoto[1], D Murayama[1], M Takeshita[1], Y Ura[1], S Abe[1], T Numazawa[2], H Takata[3], Y Matsumoto[3] and T Kuriiwa[3]

[1] Department of Physics, Kanazawa University, Kanazawa 921-8111, Japan
[2] National Institute of Materials Science, Tsukuba 302-0003, Japan
[3] Santoku Corp., Akashi 674-0093, Japan

k.matsu@staff.kanazawa-u.ac.jp

Abstract. Magnetic materials with large magnetocaloric effect are significantly important for magnetic refrigeration. $La(Fe_{0.88}Si_{0.12})_{13}$ compounds are one of the promising magnetocaloric materials that have a first order magnetic phase transition. Transition temperature of hydrogenated $La(Fe_{0.88}Si_{0.12})_{13}$ increased up to room temperature region while keeping metamagnetic transition properties. From view point of practical usage, bonded composite are very attractive and their properties are important. We made epoxy bonded $La(Fe_{0.88}Si_{0.12})_{13}$ hydrides. Magnetocaloric effect was studied by measuring specific heat, magnetization, and temperature change in adiabatic demagnetization. The composite had about 20% smaller entropy change from the hydrogenated $La(Fe_{0.88}Si_{0.12})_{13}$ powder in 2 T. Thermal conductivity of the composite was several times smaller than $La(Fe,Si)_{13}$. The small thermal conductivity was explained due to the small thermal conductivity of epoxy. Thermal conductivity was observed to be insensitive to magnetic field in 2 T. Thermal expansion and magnetostriction of the composite material were measured. The composite expanded about 0.25% when it entered into ferromagnetic phase. Magnetostriction of the composite in ferromagnetic phase was about 0.2% in 5 T and much larger than that in paramagnetic phase. The composite didn't break after about 100 times magnetic field changes in adiabatic demagnetization experiment even though it has magnetostriction.

1. Introduction

Magnetic refrigeration makes use of the magnetocaloric effect (MCE). Because of the reversibility of entropy changes, magnetic refrigeration can achieve a very high thermodynamic efficiency. Magnetic refrigeration systems can be environmentally friendly, quiet, and potentially more efficient than conventional gas expansion systems. Magnetic materials with large magnetocaloric effect are significantly important for magnetic refrigeration. Gadolinium and Gd-based alloys are the benchmark MCE material in room temperature region.

$La(Fe,Si)_{13}$ compounds are one of the promising magnetocaloric materials that have a first order magnetic transition (FOMT). This material is attractive because of large MCE and abundance. $La(Fe,Si)_{13}$ hydrides is widely studied because hydrogenation of $La(Fe,Si)_{13}$ rises the transition temperature up to room temperature region, keeping the magnetic transition as first order one. $La(Fe,Si)_{13}$ hydrides are brittle and available only in powder or small pieces and it is necessary to use

IOP Conf. Series: Journal of Physics: Conf. Series **897** (2017) 012011 doi:10.1088/1742-6596/897/1/012011

some processing to make required forms. Polymer-bonded method was tried to make composite materials from $La(Fe,Si)_{13}$ and $La(Fe,Mn,Si)_{13}H_x$ [1,2]. From view of designing of refrigerator with compositional magnetocaloric materials, thermal conductivity and other properties are also important issues.

In this study, epoxy bonded $La(Fe_{0.88}Si_{0.12})_{13}$ hydrides were fabricated and their thermal and magnetocaloric properties were investigated. Magnetocaloric effect was studied by measuring specific heat, magnetization, and temperature change in adiabatic demagnetization. Thermal conductivity, thermal expansion and magnetostriction were also measured.

2. Experiments

Bulk $La(Fe_{0.89}Si_{0.11})_{13}$ flakes were prepared from melt by a single-roll process. After homogenization process, the $La(Fe_{0.88}Si_{0.12})_{13}H_x$ material was hydrogenated. The hydrogenated powder was sieved, particle size used for epoxy composite was between 38 and 75 μm. Compacts for thermal conductivity measurement were prepared as follows; the hydrogenated powder was mixed with 3 wt% epoxy and compacted at about 0.1 GPa. Plates with size of about 20 mm length and 5 mm thick were obtained. Density of the epoxy bonded composite was determined as 6.4 g/cm^3 by Archimedes method. Figure 1 shows the photograph of the compact rectangular parallelepiped. The hydrogenated powder used for thermal expansion and magnetostriction measurements was prepared in another run so that the hydrogen composition was different from that for MCE and thermal conductivity measurements. The compact for thermal expansion and magnetostriction measurements were made using 4 wt% epoxy.

Specific heat was measured with Quantum Design PPMS in magnetic fields of 0 and 2 T. Magnetization curves and temperature dependence were measured with Quantum Design MPMS. The measurements of the thermal conductivity were performed by a steady-state method with a homemade apparatus in 0 and 2 T. The temperature gradient was measured between two points on the rod sample when defined amount of heat flux was passing through the sample. The thermal conductivity was calculated with the temperature gradient, corresponding heat flux, and cross section of the rod sample in terms of Fourier's law. Adiabatic demagnetization experiment was performed with homemade apparatus that has a superconducting magnet in persistent mode. The $La(Fe_{0.88}Si_{0.12})_{13}H_x$ compact suspended in a vacuum chamber with fine Kevler thread was pushed out and pulled into high magnetic field in several seconds keeping adiabatic condition. Temperature was measured with a Si diode temperature sensor. Thermal expansion and magnetostriction were measured using a homemade capacitive dilatometer that was installed in Quantum Design PPMS [3].

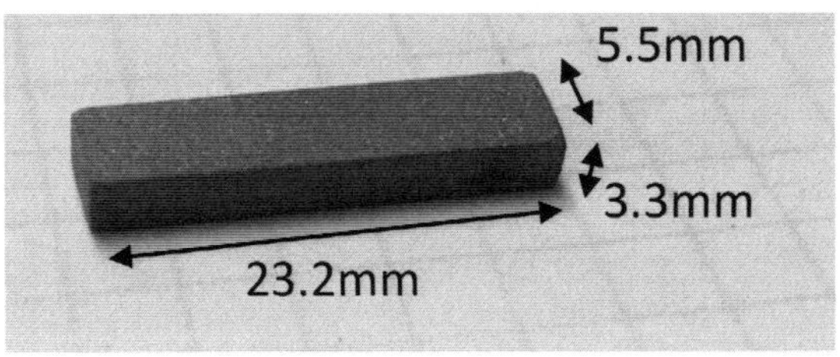

Figure 1. Photograph of the epoxy bonded $La(Fe_{0.88}Si_{0.12})_{13}$ hydrides rod for thermal conductivity measurement.

3. Results and Discussion

3.1. Specific heat

Figure 2 shows the specific heat of epoxy bonded $La(Fe_{0.88}Si_{0.12})_{13}$ hydrides in 0 and 2 T. Specific heat has a sharp peak at the ferromagnetic transition temperature, 332 K. Hydrogen concentration dependence of magnetic transition temperature for $La(Fe_{0.88}Si_{0.12})_{13}H_x$ was reported in Ref. [4]. Transition temperature increased in proportion with hydrogen concentration. From this relation, hydrogen concentration of our compact was estimated as $La(Fe_{0.88}Si_{0.12})_{13}H_{1.63}$. The peak temperature was increased by 7 K with 2 T magnetic field. The peak of specific heat is suppressed in height and broadened a little by applying magnetic field. This behaviour is typical for FOMT materials such as $Gd_5(Si_xGe_{1-x})_4$ and $La(Fe_{0.88}Si_{0.12})_{13}$. The maximum heat capacity in 0 T was near 1850 J/kgK that was close to bulk $La(Fe_{0.88}Si_{0.12})_{13}$ hydride.

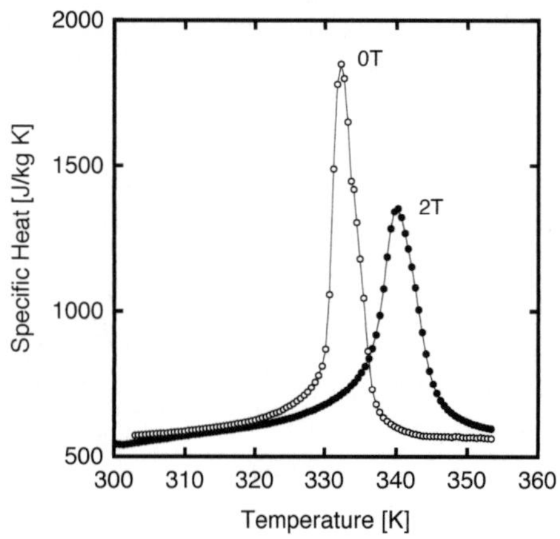

Figure 2. Specific heat of epoxy bonded $La(Fe_{0.88}Si_{0.12})_{13}$ hydrides in 0 and 2 T.

3.2. Magnetization

Magnetization of epoxy bonded $La(Fe_{0.88}Si_{0.12})_{13}$ hydrides was shown in Fig. 3 as functions of temperature. It is clearly shown in small magnetic fields that magnetization represent kinks at temperatures corresponding to FOMT. Magnetization of ferromagnetic phase saturated in about 1 T.

Magnetic entropy change (ΔS) was calculated using Maxwell's relation as $\Delta S = \int_0^H \left(\frac{\partial M}{\partial T}\right) dH$. Figure 4 shows ΔS of epoxy bonded $La(Fe_{0.88}Si_{0.12})_{13}$ hydrides as functions of temperature. The maximum $|\Delta S|$ of epoxy bonded $La(Fe_{0.88}Si_{0.12})_{13}$ hydrides was 17.2 J/kgK which is larger than that of polymer bonded $La(Fe,Mn,Si)_{13}H_x$ in Ref. [2]. The temperature region with large ΔS was expanded to higher temperature by higher magnetic field, keeping the maximum ΔS as almost constant. This behaviour of ΔS with various magnetic fields is often observed in FOMT materials.

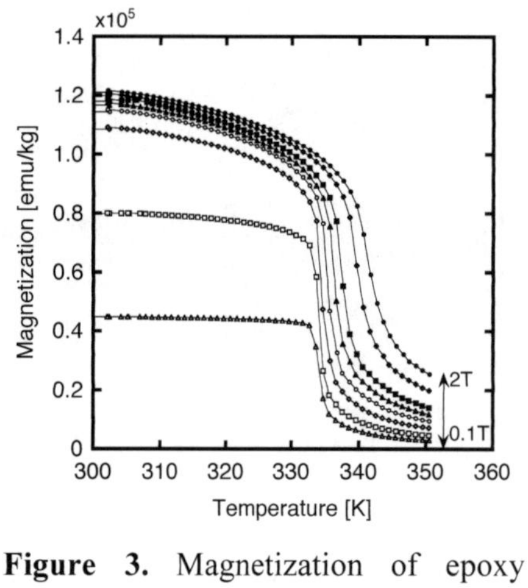

Figure 3. Magnetization of epoxy bonded $La(Fe_{0.88}Si_{0.12})_{13}$ hydrides in 0.1, 0.2, 0.4, 0.6, 0.8, 1, 1.5, and 2 T.

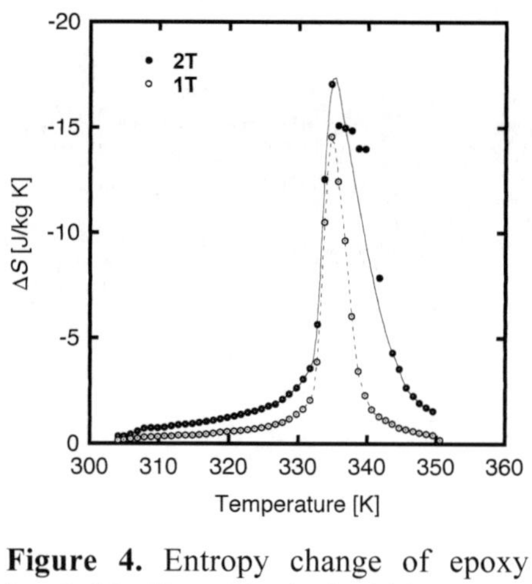

Figure 4. Entropy change of epoxy bonded $La(Fe_{0.88}Si_{0.12})_{13}$ hydrides

3.3. Entropy-Temperature diagram

The total entropy (S) was obtained by integrating specific heat divided by temperature (C/T) with respect to temperature at constant magnetic field, as $S(T,H) = \int_0^T \frac{C(T,H)}{T} dT$, where $C(T, H)$ is the specific heat at a constant field H. The $S(T, H)$ was determined so that ΔS obtained from the specific heat and the ΔS obtained from magnetization were consistent with each other. Figure 5 gives the entropy-temperature (S–T) diagram of epoxy bonded $La(Fe_{0.88}Si_{0.12})_{13}$ hydrides. Entropy increases rapidly at the transition temperature because of FOMT.

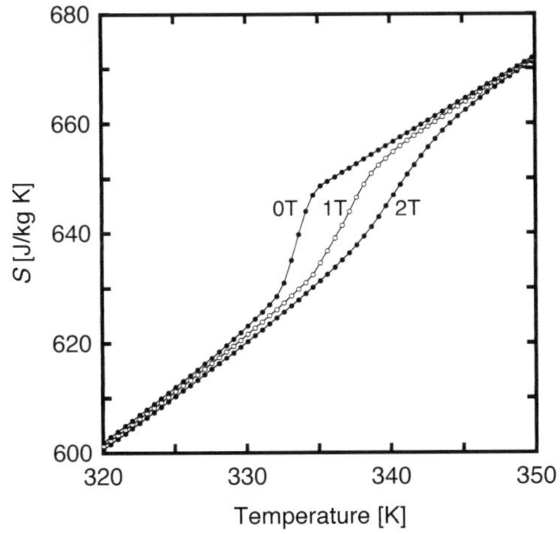

Figure 5. Entropy-Temperature diagram of epoxy bonded $La(Fe_{0.88}Si_{0.12})_{13}$ hydrides in 0, 1, and 2 T.

3.4. Temperature change in Adiabatic demagnetization

Figure 6 represents the temperature change (ΔT) in adiabatic demagnetization and magnetization. It was confirmed that temperature changes almost coincided in magnetization and demagnetization processes (not shown in Fig. 6). Abscissa of Fig. 6 represents the starting temperature of demagnetization. Maximum temperature changes of 3.5 and 5.5 K were obtained at 337 and 339 K with 1 and 2 T, respectively. Those temperature changes were in good agreement with those estimated from the S–T diagram. The composite maintained its shape after about 100 times magnetic field changes in adiabatic demagnetization experiment even though it has magnetostriction as explained in section 3.6.

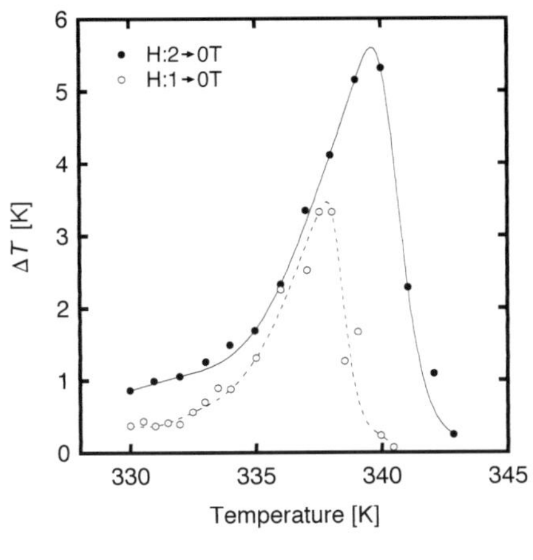

Figure 6. Adiabatic temperature change of epoxy bonded $La(Fe_{0.88}Si_{0.12})_{13}$ hydrides . Abscissa represents the starting temperature of demagnetization.

IOP Conf. Series: Journal of Physics: Conf. Series **897** (2017) 012011 doi:10.1088/1742-6596/897/1/012011

3.5. Thermal conductivity

In order to discuss heat transfer between fluid and magnetic material, thermal conductivity of epoxy bonded compact is important. Figure 7 shows thermal conductivity of epoxy bonded $La(Fe_{0.88}Si_{0.12})_{13}$ hydrides in 0 T. The difference of thermal conductivity in 0 and 2 T was within our experimental error. As shown in the Fig. 7, thermal conductivity of our epoxy bonded compact is about three times smaller than that of bulk [5]. The thermal conductivity due to electron was estimated as 60% of total κ using Wiedeman-Franz law and electric conductivity for $La(Fe_{0.88}Si_{0.12})_{13}H_{0.4}$ [6]. It looks to be reasonable that our epoxy bonded compact had smaller thermal conductivity than that of bulk. Because epoxy is an insulator and the thermal conductivity of epoxy is about two order of magnitude smaller than $La(Fe_{0.88}Si_{0.12})_{13}$, 3 wt% epoxy caused significant decrease in thermal conductivity of composite material. Radulov et al. made $La(Fe_{0.88}Si_{0.12})_{13}$ hydrides polymer composite in various processes and measured thermal conductivity [2]. Thermal conductivity of the compact was ~5 W/Km at around 300 K. Low thermal conductivity of our compact might be attributed to inclusion of voids so that detailed study on synthesis pressure, particle size, and adhesion are necessary to improve thermal conductivity of epoxy bonded magnetocaloric material.

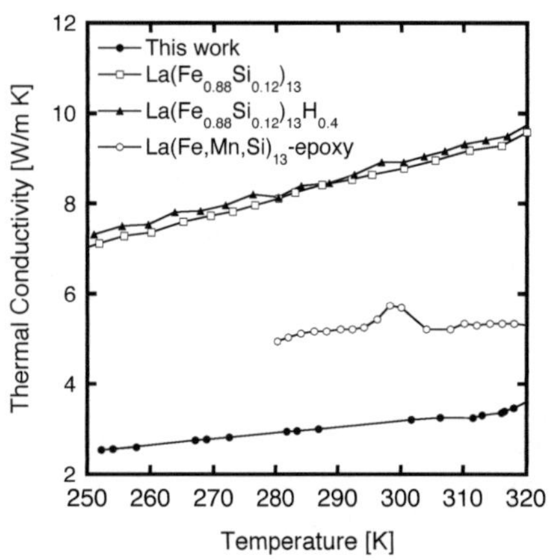

Figure 7. Thermal conductivity of epoxy bonded $La(Fe_{0.88}Si_{0.12})_{13}$ hydrides in 0 and 2 T. That of bulk $La(Fe_{0.88}Si_{0.12})_{13}$ is plotted form literatures [2, 6]

3.6. Thermal expansion and magnetostriction

Our epoxy bonded $La(Fe_{0.88}Si_{0.12})_{13}$ hydrides expanded about 0.25% when it entered into ordered phase in 0 T as shown in Fig. 8. Thermal hysteresis was not significant. The expansion ($\Delta L/L$) of bulk $La(Fe_{0.88}Si_{0.12})_{13}$ was measured as about 0.3% at the transition by X-ray diffraction [7]. $\Delta L/L$ of our compact did not change so rapidly as bulk. Epoxy possibly made abrupt volume change smooth.

Figure 9 shows magnetostriction of epoxy bonded $La(Fe_{0.88}Si_{0.12})_{13}$ hydrides as functions of squared magnetic field in ferromagnetic phase, around transition temperature, and paramagnetic phase. In all temperatures, the epoxy bonded compact expands by magnetic field. Hysteresis wasn't observed. Large magnetostriction was observed only around transition temperature and rapidly changed from 0 to 2 T. In paramagnetic phase, magnetstriction is in proportion to squared field. At well below transition temperature, magnetostriction shows saturation. The composite maintained its shape after thermal expansion and magnetostriction measurements.

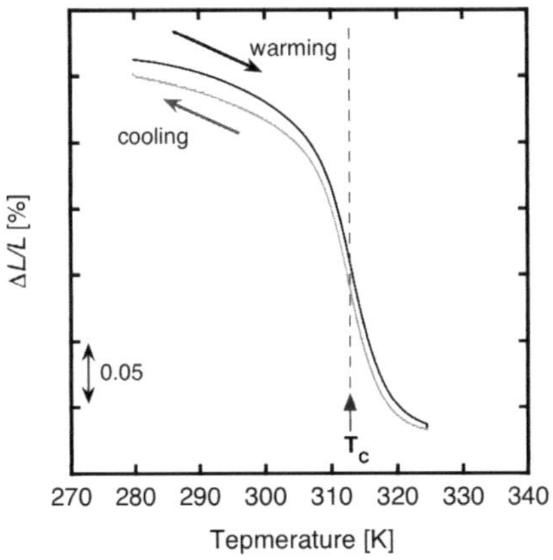

Figure 8. Thermal expansion of epoxy bonded $La(Fe_{0.88}Si_{0.12})_{13}$ hydrides in 0 T.

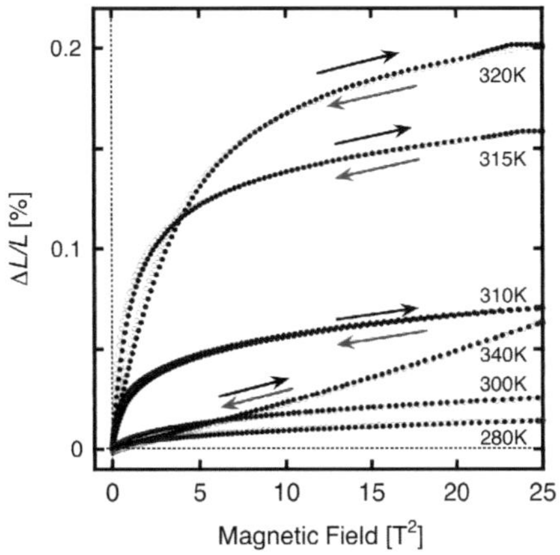

Figure 9. Magnetostriction of epoxy bonded $La(Fe_{0.88}Si_{0.12})_{13}$ hydrides. Magnetostriction is plotted as functions of squared magnetic field.

4. Summary

We made epoxy bonded $La(Fe_{0.88}Si_{0.12})_{13}$ hydrides. Magnetocaloric effect was studied by measuring specific heat, magnetization, and temperature change in adiabatic demagnetization. The epoxy bonded compact maintained sharp magnetic transition so that have large magnetocaloric effect. Entropy-temperature diagram was obtained from these measurements. Adiabatic temperature change in adiabatic demagnetization was observed and coincided with that estimated from entropy-temperature diagram. Thermal conductivity was several times smaller than bulk $La(Fe_{0.88}Si_{0.12})_{13}$ due to low thermal conductivity of epoxy adhesion. The compact expanded 0.25% when it entered into ferromagnetic phase. Magnetostriction was largest around the transition temperature.

5. Acknowledgments

This work is supported by JSPS KAKENHI Grant Number 26420720 and 26420890.

References

[1] Skokov K P, Karpenkov D Yu, Kuz'min M D, Radulov I A, Gottschall T, Kaeswurm B, Fries M and O. Gutfleisch 2014 *J. Appl. Phys.* **115** 17A941
[2] Radulov I A, Skokov K P, Karpenkov D Y, Gottschall T and Gutfleisch O 2015 *J. Mag. Mag. Mater.* **396** 228
[3] Iwakami O, Kawata N, Takeshita M, Yao Y, Abe S and Matsumoto K 2014 *J. Phys.: Conference Series* **568** 032002
[4] Fujita A, Fujieda S, Hasegawa Y and Fukamichi K 2003 *Phys. Rev.* B **67** 104416
[5] Fujieda S, Hasegawa Y, Fujita A and Fukamichi K 2004 *J. Appl. Phys.* **95** 2429
[6] Fukamichi K, Fujita A and Fujieda S 2006 *J. Alloys and Compound* **408-412** 307
[7] Fujita A, Fujieda S, Fukamichi K, Mitamura H and Goto T 2001 *Phys. Rev.* B **65** 014410

Cryogenic Adsorption of Nitrogen and Carbon Dioxide in Activated Carbon

Fuzhi Shen[1,2], Huiming Liu[1], Dong Xu[1], Hengcheng Zhang[1,2], Junfeng Lu[1] and Laifeng Li[1,2]
[1]State Key Laboratory of Technologies in Space Cryogenic Propellants, Technical Institute of Physics and Chemistry, Chinese Academy of Sciences, Beijing 100190, P. R. China.
[2]University of Chinese Academy of Sciences, Beijing 100049, P. R. China

Email: huimingliu@mail.ipc.ac.cn

Abstract. Activated carbon have been used for a long time at low temperature for cryogenic applications. The knowledge of adsorption characteristics of activated carbon at cryogenic temperature is essential for some specific applications. However, such experimental data are very scare in the literature. In order to measure the adsorption characteristics of activated carbon under variable cryogenic temperatures, an adsorption measurement device was presented. The experiment system is based on the commercially available PCT-pro adsorption analyzer coupled to a two-stage Gifford McMahon refrigerator, which allows the sample to be cooled to 4.2K. Cryogenic environment can be maintained steadily without the cryogenic liquid through the cryocooler and temperature can be controlled precisely between 5K and 300K by the temperature controller. Adsorption measurements were performed in activated carbon for carbon dioxide and nitrogen and the adsorption isotherm were obtained.

1. Introduction

All gases can be condensed easily except helium and hydrogen at cryogenic temperature of about 4K. The cryosorption pumps used for creating high vacuum for these gases are characterized by high pumping speed, brilliant storage capacity and low end pressure. [1] Adsorption isotherms are strongly needed for the design of these kind of applications. However, such experiment data are very scare in the literatures. [2]

Resulting from the high porosity, activated carbons are characterized by large inner pore surface area. For this reason, activated carbons are widely used in the integration of cryogenic devices for their adsorption properties, especially in the pumping of gases such as helium and hydrogen and their isotopes for some special applications. [3,4]

An adsorption measurement device was built in this paper which allow us systematically measure the adsorption characteristics of chosen adsorbent-adsorbate pairs. The experiment system is based on the commercially available PCT-pro adsorption analyzer coupled to a two-stage Gifford McMahon refrigerator, which allows the sample to be cooled to 4.2K. Nitrogen and mercury porosimetry facilities were used for testing the characterization of the samples. The adsorption isotherms are obtained by home built adsorption measurement device, including adsorption isotherms for CO_2 at 300K and 217K and N_2 at 220K and 170K.

IOP Conf. Series: Journal of Physics: Conf. Series **897** (2017) 012012 doi:10.1088/1742-6596/897/1/012012

2. Experimental Setup

2.1 Carbon Dioxide

Carbon dioxide has been widely used in cryogenic application for its large inner pore surface. For this reason, carbon dioxide has been selected as sorbent in this paper. In consideration of its high adsorption capacity, high hardness and wide grain size range, we choose PICACTIF TA 60, a kind of regenerable granular microporous activated carbon, as sample used in our experiment. Nitrogen and mercury porosimetry facilities were used for testing the characterization of the samples. Table 1 presents the porosity characterization of the activated carbon used in experiment.

Table 1. Porosity characterization of the activated carbon used in experiment.

surface area(m^2/g)	1376.91
pore volume(cc/g)	0.547
apparent density(g/cm^3)	0.45-0.53
Humidity as packed(% min)	5
CCl4 index(% min)	60-70

2.2 Gases used

As BET has become a recognized way to establish standards in the characterization of porous materials, experiments with nitrogen were carried out in our experiment. [5] With more attentions have been paid to the green house effects and their impacts, carbon dioxide were used as adsorptives in this paper, too. The nitrogen and carbon dioxide gases were used in a purity of 99.9999 vol.%.

The core parameters used in this paper were listed in table 2.

Table 2. Selected data of nitrogen and carbon dioxide

Gas	T_c[K]	p_c[kPa]	p_0[kPa](T_0[k])	M[g/mol]
N_2	126.19	3397.80	101.35(77.35)	28.0134
CO_2	304.41	7386.60	53.065(217)	44.0095

2.3 Experiment principle

The adsorption measurement device mentioned in this paper use a 'Sieverts apparatus' way to measure the solubility of gases in the sample. [6]

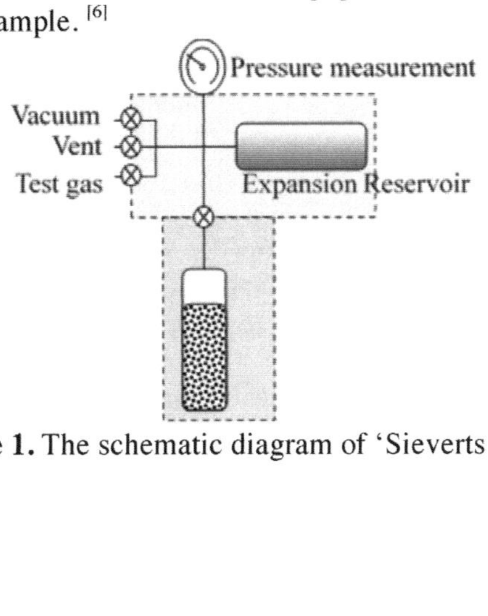

Figure 1. The schematic diagram of 'Sieverts apparatus'

When we use 'Sieverts apparatus' to measure the adsorption characteristics, sample at known pressure and volume will be connected to a reservoir volume and pressure through an isolation valve. When value is closed, reservoir is filled and equilibrated. Then we open the isolation value, gases will diffuse until the new equilibrium establish. Sample pressure is equilibrated and measure in this moment. Through this way, gas adsorption can be calculated by formula 1 based on the conservation of mass.

$$\Delta n = \left(\frac{P_{s(m-1)} \cdot V_s}{Z_s \cdot R \cdot T_s} + \frac{P_{r(m)} \cdot V_s}{Z_r \cdot R \cdot T_r}\right) - \left(\frac{P_{s(m)} \cdot V_s}{Z_s \cdot R \cdot T_s} + \frac{P_{s(m)} \cdot V_r}{Z_r \cdot R \cdot T_r}\right) \qquad (1)$$

- n: the amount of gas adsorption
- $Ps_{(m-1)}$: the pressure of the sample reservoir in (m-1) moment
- $Pr_{(m-1)}$: the pressure of the expansion reservoir in (m-1) moment
- Vs: the volume of the sample reservoir
- Vr: the volume of the expansion reservoir
- Ts: the temperature of the sample reservoir
- Tr: the temperature of the expansion reservoir
- Zs: the compressibility factor of gas in the sample reservoir
- Zr: the compressibility factor of gas in the expansion reservoir

2.4 Experimental Setup

The adsorption measurement device mentioned in this paper is made up of three main parts: a commercially available PCT-pro adsorption analyzer, a sample chamber to accept the samples and a two-stage Gifford McMahon refrigerator for creating cryogenic testing environment. With the refrigerator unit integrated with the PCT-pro adsorption analyzer, we can study the sample to the temperature between 4.2K and 300K.

PCT-pro adsorption analyzer is a fully automated Sieverts instrument for measuring gas adsorption properties of materials. It is designed for precision measurements over a broad range of pressures, temperatures, and sample sizes. This instrument hosts a number of advanced features including PID controlled gas pressure, temperature regulated gas handling system, different calibrated volumes, automatic high and low-range pressure switching and sophisticated measurement and process.

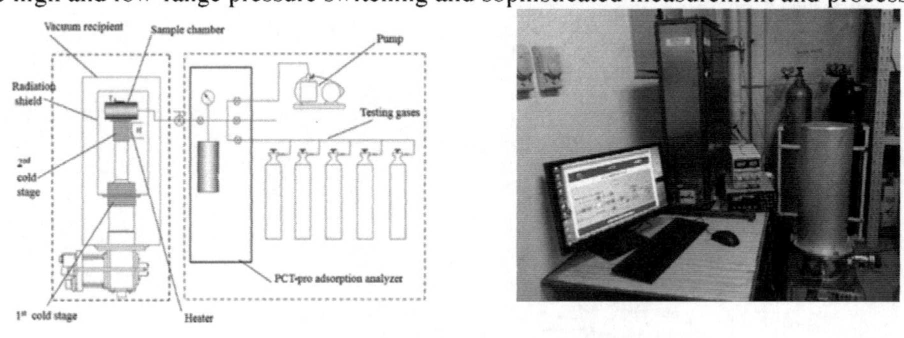

Figure 2. Home built adsorption measurement device

Instead of using liquid nitrogen (77K as a minimum) or liquid helium (4.2K as a minimum) to cool the sample, the home built adsorption measurement device mentioned in this paper use a Gifford McMahon refrigerator as the cooling source. The refrigerator includes three main parts: a two-stage cold head, two transfer line and a compressor.

The sample chamber was installed on the cold head and connected to the PCT-pro adsorption analyzer by means of gas transfer line.

The cold head can attain temperature down to 30K at the first stage and 4.2K at the second stage based on the heat input. A radiation shield was installed in the first stage to reduce the thermal load of the second stage. A vacuum recipient was installed to maintain the vacuum environment. In order to control the testing temperature concisely, we choose a temperature controller made by model 32B to adjust the temperature and 150W heater was attached to the second stage.

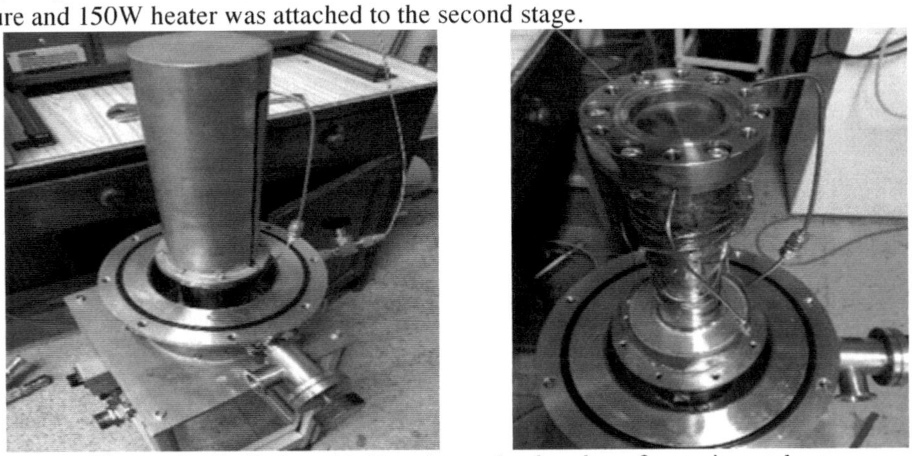

Figure 3. Radiation shield and sample chamber of experimental setup

3. Experimental Procedure and results

3.1 Experimental Procedure

The whole adsorption testing experiment can be divided into three main parts: preparation of the sample, set of the testing environment and measurement of the sample.

In the first step of preparing the sample, we should put the sample into the vacuum oven and heat the sample in 400K at least 24 hours. After the sample being totally heated and vacuum, the sample should be weighed.

There are two environment condition we should set up to guarantee the experiment to run well: vacuum condition and temperature condition. In order to creating vacuum environment, we should put a copper shim into the gap between the two side of the chamber room after put the sample into the sample chamber room, then close the sample chamber room, the radiation shield and the vacuum recipient. Immediately afterwards, start to vacuum the vacuum recipient and the radiation shield at least 24 hours by the vacuum pump. After the vacuum condition is reached, the cryocooler is turned on and temperature controller can start to work to acquire the temperature condition.

Until the temperatures are reached equilibrium in the sample chamber, the radiation shield and the cold head, we can start the adsorption measurement experiment.

The first step of adsorption measurement experiment is evacuating the whole system, then calibrating the volume of the sample chamber. After the volume was obtained, the appropriate point on the menu can be used to start the adsorption measurement experiment. In this time, we should input the parameter we need to the computer. Then, the experiment start and the data can be recorded by the computer automatically.

3.2Experiment Results

The adsorption measurements were carried out with CO_2 and N_2. Adsorption isotherms were obtained for CO_2 as adsorbate at 300K and 220K and nitrogen as adsorbate at 217K and 170K. From figure 4 and figure 5, we can tell the isotherms of CO_2 belong to Type-I adsorption isotherm and the isotherms of N_2 are Type-II isotherms according to IUPAC.

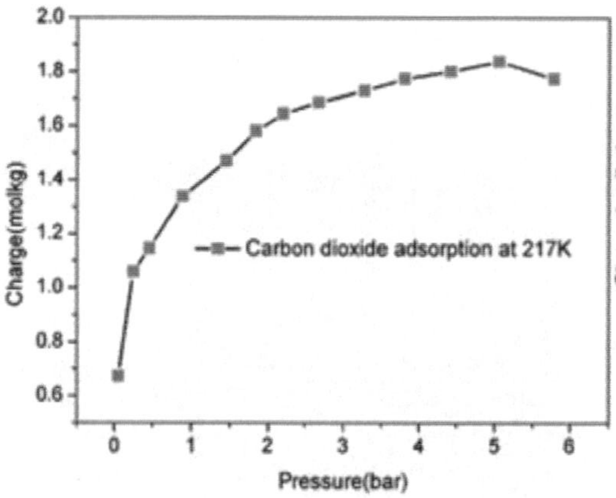

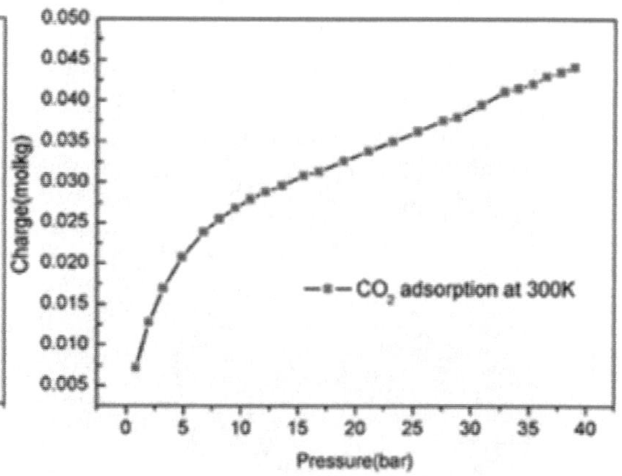

Figure 4. Adsorption isotherm for Carbon dioxide at 300K and 200K

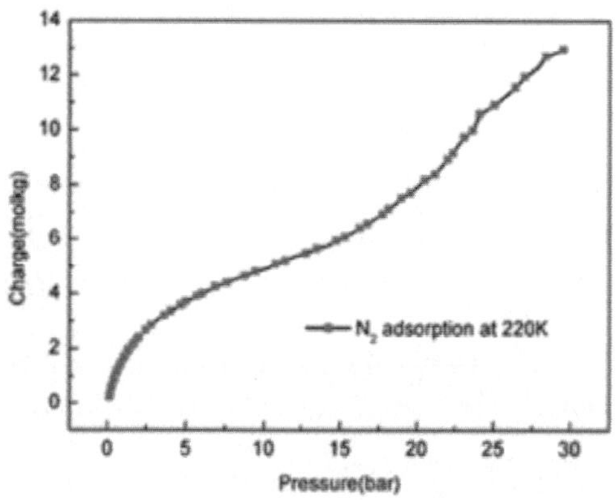

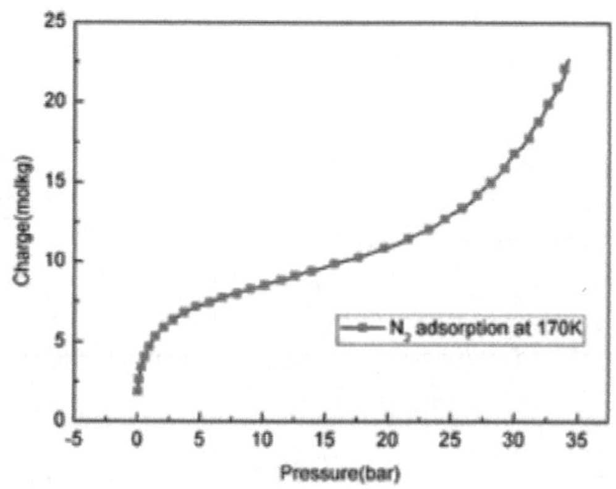

Figure 5. Adsorption isotherm for Nitrogen at 300K and 200K

4. Conclusion

An adsorption measurement device has been designed and built in a temperature range of 4.2K to 300K. The cryogenic adsorption measurements were carried out with CO_2 and N_2 as adsorbate. The porosity characterization of the activated carbon and the core parameters of gases used in the experiment were studied. Adsorption isotherms were obtained for CO_2 and N_2 at specific temperature.

References

[1] V. Hauer, Chr. Day, Cryosorbent Characterization of Activated Charcoal in the COOLSORP Facility, 2002

[2] U. Behera, et al., Develooment of experimental setup for study of adsorption characteristic of porous materials down to 4.2K. Advances in Cryogenic Engineering, 2012, p1285-1292

[3] D.Martins, I. Catarino, D.Lopes, I. Esteves, J.P. Mota and G. Bonfait, Low temperature Adsorption Versus Pore Size in Activated Carbons, International Crycoller Conference, 2011, p567-573

[4] F. Rouquerol, J.Rouquerol, K. S. W. Sing, Adsorption by Powders and Porous Solids, Academic Press, London, 1999

[5] Duband,L., Collaudin,B., "Sorption Coolers Develpoment at CEA_SBT," Cryogenics, Vol. 39, Issue:8, August 1999, p659-663.
[6] D. D. Do, Adsorption Analysis: Equilibria and Kinetics, Imperial College Press,1998

Acknowledgements

This work was supported by the National Natural Science Foundation of China (No.51406217), and the fund of the state key laboratory of technologies in space cryogenic propellants, SKLTCP1601

IOP Conf. Series: Journal of Physics: Conf. Series **897** (2017) 012013 doi:10.1088/1742-6596/897/1/012013

The first results of He II boiling visualization experiment conducted under 4.7 seconds microgravity conditions

S Takada[1], N Kimura[2], S Pietrowicz[3], K Grunt[3]

[1] National Institute for Fusion Science, 322-6 Oroshi cho, Toki, Gifu,509-5292 Japan
[2] High Energy Accelarator Research Organization, 1-1 Oho, Tsukuba, Ibaraki, Japan
[3] Wroclaw University of Science and Technology, Wybrzeze Wyspianskiego 20, 50-370 Wroclaw, Poland

takada.suguru@LHD.nifs.ac.jp

Abstract. An experiment regarding boiling of superfluid helium (He II) has been carried out under conditions of microgravity, in order to investigate the dynamics of the phase transition. A small cryostat equipped with visualization setup has been utilized for this purpose. Presence of two orthogonal optical axes allowed for registering of 3-dimensional images of a vapor bubble induced by a micro heater. Microgravity environment has been produced for about 4.7 s using the 122 m high drop tower facility at ZARM (Center of Applied Space Technology and Microgravity), University of Bremen, Germany. The experimental campaign consisting of 32 drops has been successfully conducted, while avoiding any damage to the equipment.

1. Introduction

He II is utilized as a coolant in satellite telescopes such as X ray telescopes and infrared telescopes. Thus, it is important to become acquainted with the heat transfer mechanisms in He II under microgravity. However, only a few experiments regarding this matter have been conducted so far [1, 2]. The knowledge of the boiling phenomena in He II under microgravity conditions is necessary for a proper design of space equipment exposed to relatively high heat loads.

Furthermore, it is well known that the superfluid helium heat transfer accompanied by boiling depends on the subcooling caused by the presence of a hydrostatic pressure head. Thus, an experimental study conducted under the gravity of Earth would not be representative for a microgravity environment. Moreover, a microgravity experiment may lead to revealing the mechanisms of heat transfer in He II otherwise overshadowed by the influence of gravitational field during a ground experiment.

A series of microgravity experiments has been conducted at the drop tower of Hokkaido Center of AIST over the course of several years [3-5]. The drop tower of AIST can produce microgravity conditions only for about 1.3 s. In this study, the visualization of bubble growth was aimed at reaching a steady state, what was facilitated by longer a microgravity period of about 4.7 s.

2. Experimental set-up

2.1. Drop tower

The microgravity environment was produced using the drop tower of ZARM. The drops have been performed inside a 122 m high vacuum tube in order to ensure high quality of microgravity. Vaccuum pressure inside the drop tube was being reduced to approximately 50 Pa, ensuring that the residual gravity drops below 10^{-4} *g* (1 *g* equals the value of the standard acceleration due to gravity on ground)[6]. The duration of microgravity was abut 4.7 s. The first 0.5 s just after the beginning of free fall were not considered, because of the residual acceleration due to initial vibrations. The experimental set-up as shown in Fig.1 was installed in an airtight capsule. Power for all of the experimental set-up components was supplied from a battery pack. The dropped capsule was being captured by a reciever tank filled with vast powder beads, which results in deceleration ranging up to 45 *g*

2.2 *Cryostat*

2.2-1 Safety measures

The utilized cryostat was modified in comparison with previous research [4, 5] to withstand the deceleration of 45 *g*. Its weakest point was the welding seal joint on the top of the neck, which may cause problems during deceleration for vertical axis. Hence, the copper radiation shield was replaced with aluminium one to reduce the weight. In addition, several G-10 supports were set to fix the lateral axis. The safety factor assumed at the design stage always exceeded 2.

The mitigation of a false safe was considered. When a sudden vacuum loss happens, harmful boiling of liquid helium occurs inside the vessel. The evaporated gas is vented out of the capsule using a feedback operation of the control valve with a pressure sensor on the vent line. In addition, numerous safety valves were set in the capsule in order to mitigate the results of the worst-case scenario when the valve control doesn't work. On top of the cryostat a weak safety valve which opens just above atmospheric pressure was installed. In addition two safety valves were set on the boundary between the drop capsule and the vacuum drop tube. The schematic illustration of the cryostat and the valves is shown in Fig 2.

Fig.1 A picture of the experimental set-up inside the capsule

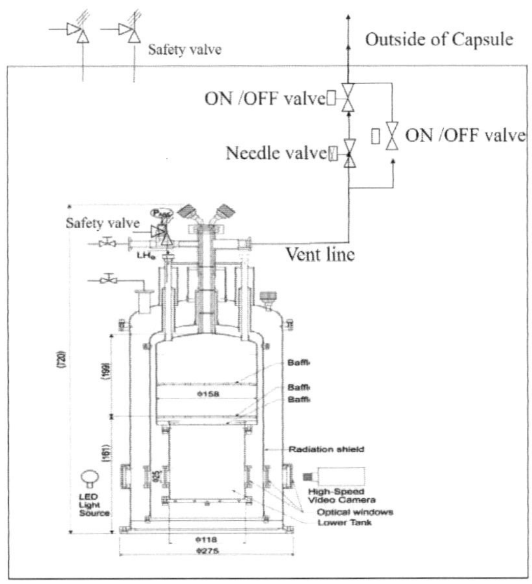

Fig.2 The schematic illustration of the cryostat and the valves

2.2-2 Experimental set-up and operation

The cryostat was equipped with 4 windows for visualization over two orthogonal optical axes. Two LED arrays and two small high speed cameras (Photoron Fastcam MC-2) along with a pair of telecentric lens were set to obtain a 3-dimensional image of vapor bubble boundary position. A small heater which consisting of a short Manganin wire (φ0.05 x 1.88 mm) and two superconducting wires [4, 5] was placed vertically. Consequently the axis of the heater is perpendicular to the plane defined by the optical axes. The heater current was supplied from a constant power supply utilizing a feedback control. The data acquisition system and all of the solenoid valves were controlled using the capsule control system provided by ZARM, which is based on Labview® and National Instruments PXI® platform.

Liquid helium was transferred before each drop prior to enclosing the capsule and launching it within the drop tube. The test vessel was pumped out using the large vacuum volume of the drop tube via the vent line across the boundary of the capsule. Prior to release of the capsule, the pressure inside the cryostat was controlled using the electric valve on the vent line. During the free fall period, all of the valves were closed to avoid a thruster effect to the capsule. If the tiny exhaust leakage exists between capsule and vacuum tower, the falling capsule must be tilted by thruster effect. Thus, the temperature of He II inside the cryostat was increasing by about several mK during the free fall. For the purpose of this study, this relatively small temperature increase may be considered negligible.

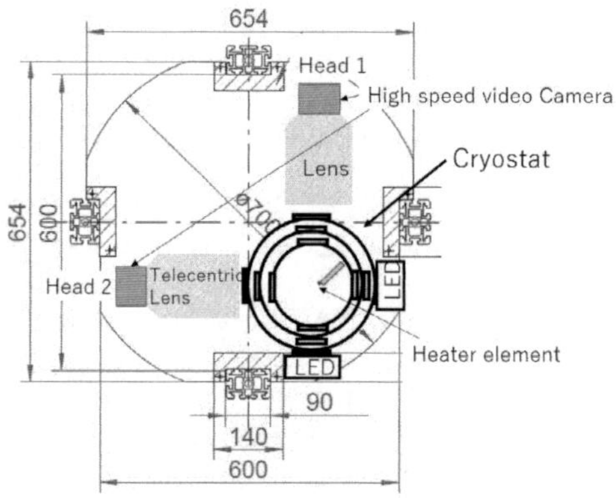

Fig. 3 A top view of the experimental set-up

3. Results

All of the 32 scheduled drops were successfully conducted. The experimental set-up withstood decelerations ranging up to 45 *g*. Typical visualization results of He II boiling under microgravity are shown in Fig. 4. Fig. 4 (a) presents an image from the camera head 1, whereas Fig. 4 (b) corresponds to the camera head 2. The shape of the vapor bubble is very close to an exact sphere. The center of the bubble is shifted from the center of wire. These characteristic features were always observed, regardless of the bulk temperature or dissipated heat flux.

Typical time distributions of the heater power and temperature are shown in Fig. 5 and Fig. 6, respectively. In this paper, the time $t = 0$ defines the beginning of the free fall. The heater current was applied after 0.5 s from free fall start. The dissipated power was controlled within about 2% by the feedback system, though the electric resistance of the Manganin wire varied during heating. Fig. 6 shows the calculated temperature of the heater wire, computed on the basis of its resistance. This temperature seems to oscillate depending on the distance between the heater wire and the vapor-liquid interface.

The size of the bubbles generated at 1.9 K was compared with the values previously obtained during experiments held at Hokkaido center of AIST, as shown in Fig.7. The duration of microgravity in the previous experiments was about 1.3 s. In the previous study, the longest heating duration under microgravity condition is 1.2 s, because residual gravity is rather high during the initial 0.1 s of a drop.

The obtained values and the tendency of the present experimental results are consistent with the previous results. Thus, it can be concluded that repeatability was confirmed. In the case of the longest heating duration of 4.2 s, the bubble growths in between the bubbles for the case of 1.2 s and the calculated value of simple model of steady heat transfer across the liquid-vapor interface. The calculated steady-state bubble diameters are derived from the energy balance on the vapor-liquid interface, based on kinetic theory [7-9]:

$$2\sqrt{\pi}\left(\frac{1-0.4\beta}{\beta}\right)j-\Delta p\sqrt{2RT_i}+\frac{\sqrt{\pi}}{4}q_i=0 \quad (1)$$

where j denotes the mass flux across the interface, β is the thermal expansion ratio, q_i is the heat flux across the interface, and T_i is the temperature at the interface (not the bath temperature). $T_i = T_b$ can be assumed because the temperature difference between them is negligibly small. T_i -T_b is caused by the temperature distribution of He II and the thermal resistance between He II and He Vapor. The thermal resistance across He II-vapor interface are quite small [10, 11]. And temperature distribution should be close to zero because of zero subcooling condition realized in microgravity. Only the pressure difference due to surface tension cause slight subcooling. Even if several ten-micron bubble appears in initial phase, the possible maximum temperature difference is the order of 10 mK. Thus, the temperature difference must influence negligibly small against the calculation results. Under microgravity conditions, the pressure difference at the interface is attributed exclusively to surface tension.

During the steady state the mass flux j equals 0, hence the first term of Eq. (1) is excluded and q_i is obtained easily from the following expression:

$$q_i=\frac{4}{\sqrt{\pi}}\left(\frac{4\sigma}{D}\sqrt{2RT_b}\right) \quad (2)$$

where D stands for the diameter and s refers to the surface tension. The heat flux across the bubble surface can be expressed as follows:

$$Q=q_iD^2\pi \quad (3)$$

Hence, the diameter in a steady state equals:

$$D=Q/16\sigma\sqrt{2RT_b\pi}\ . \quad (4)$$

Since a steady state was closely approached during the 4.2 s long drops, the above formula brings valid results. In the case of a small heat current, just above the critical value of boiling onset, the bubble size doesn't exceed the calculated one.

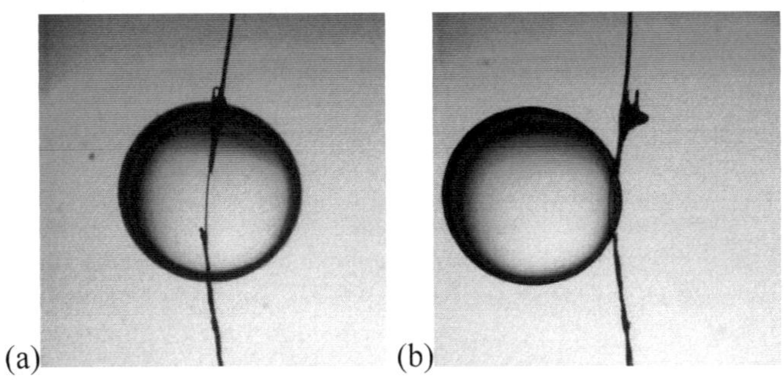

Fig. 4 Visualization results of a vapor bubble induced by the small heater; T_b = 1.9 K, q = 5.08 mW, t = 4.2 s. (a) Head 1 (b) Head 2. Visualized area is about 12.5 x 12.5 mm square.

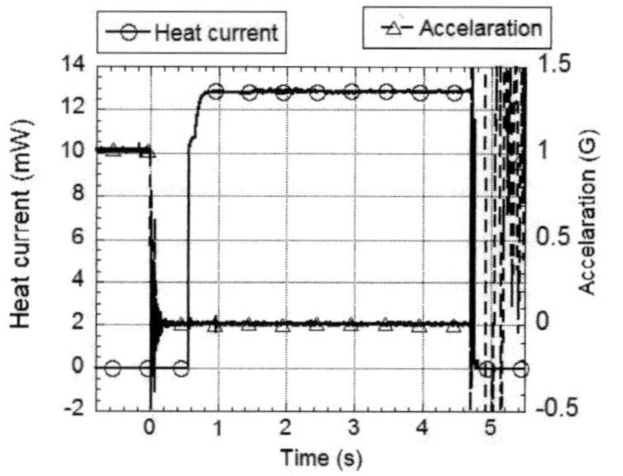

Fig.5 Time distribution of heat current and acceleration T_b = 1.9 K, q = 12.83 mW

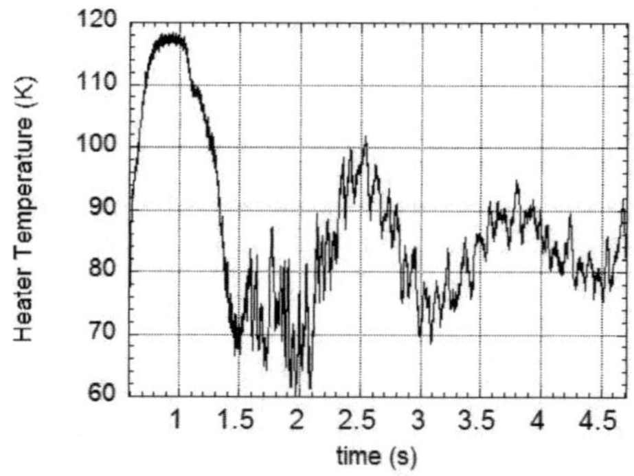

Fig.6 Time distribution of heater temperature T_b = 1.9 K, q = 12.83 mW

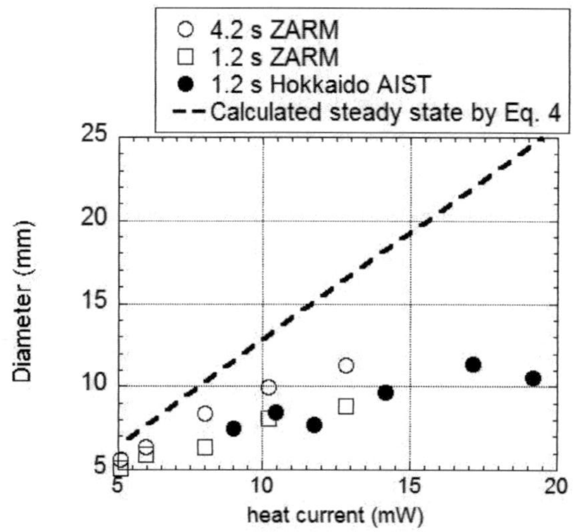

Fig.7 Bubble diameter as a function of heat current at bulk temperature equal 1.9 K

4. Summary

A series of 32 drops was successfully performed without any damage to the experimental set-up, using the 122 m high drop tower located in ZARM, Bremen, Germany. The center of the bubbles generated through boiling of saturated superfluid helium is always shifted away from the center of the heater. A complicated temperature distribution might exist inside a bubble, although due to the almost spherical shape of a bubble the heat flux across the interface can be assumed to be uniform. Repeatability of the performed experiment was confirmed through comparison with the previous results obtained at Hokkaido Center of AIST.

Acknowledgements

This study was supported by JSPS KAKENHI Grant-in-Aid for Scientific Research (B) 25289300, the grant of JAXA's study for strategic development and by European Space Agency HRE/UP/2016-01/AO.

The operation of the present experiments was supported by the team of Zentrum fur Angewaudte Raumfahrtte chnologie und Mikrogravitatation (ZARM) FAB mbH. We would like to thank especially Dr. Thorben Konemann, Mr. Jan Siemer and Mr. Thorsten Lutz. We also thank Mr. Shuichi Goto (Jecc Torisha) for his help in development of the small cryostat. Finally, we appreciate Prof. Masahide Murakami (University of Tsukuba) and Dr. Takahiro Okamura (KEK) for the fruitful discussion.

References

[1] Lipa, J.A., Swanson, D. R., Nissen J. A., Chui, T. C. P., 1994, *Cryogenics* **34** 341-347
[2] Gradt T., Szücs Z. Denne H. D. Klipping G., 1986, *Advances in Cryogenic Engineering.* **31** 499-504
[3] Kimura N., Takada S. et al, 2011, *Cryogenics* **51** 74-78
[4] Takada S., Kimura N., Murakami M., Okamura T., 2015, *Physics Procedia* **67** 591–595
[5] Takada S., Kimura N., Murakami M., Okamura T., 2015, *IOP Conference Series: Materials Science and Engineering* **101** 12163-12172
[6] ZARM Drop Tower Bremen Users Manual
[7] Ametitov Y., 1983, *Cryogenics* **23** 179-184
[8] Kryukov A. P.,van Sciver S. W., 1981, *Cryogenics* **21** 525-528
[9] Labuntzov D. A., Ametitov Y., 1979, *Cryogenics* **19** 401-404
[10] Jebali F., Francois M.X., 1990, Cryogenics 30, 355-359
[11] Jebali F., Francois M.X., 1994, Pysica B 194-196, 599-600

IOP Conf. Series: Journal of Physics: Conf. Series **897** (2017) 012014 doi:10.1088/1742-6596/897/1/012014

Design of High Voltage Electrical Breakdown Strength measuring system at 1.8K with a G-M cryocooler

Jian Li[1, 2], Rongjin Huang [1, 2*], Xu Li[1, 2], Dong Xu[1], Huiming Liu[1] and Laifeng Li[1, 2]

[1]*State Key Laboratory of Technologies in Space Cryogenic Propellants, Technical Institute of Physics and Chemistry, Chinese Academy of Sciences, Beijing, China*
[2]*University of Chinese Academy of Sciences, Beijing, China*

E-mail: huangrongjin@mail.ipc.ac.cn.

Abstract. Impregnating resins as electrical insulation materials for use in ITER magnets and feeder system are required to be radiation stable, good mechanical performance and high voltage electrical breakdown strength. In present ITER project, the breakdown strength need over 30 kV/mm, for future DEMO reactor, it will be greater than this value. In order to develop good property insulation materials to satisfy the requirements of future fusion reactor, high voltage breakdown strength measurement system at low temperature is necessary. In this paper, we will introduce our work on the design of this system. This measuring system has two parts: one is an electrical supply system which provides the high voltage from a high voltage power between two electrodes; the other is a cooling system which consists of a G-M cryocooler, a superfluid chamber and a heat switch. The two stage G-M cryocooler pre-cool down the system to 4K, the superfluid helium pot is used for a container to depress the helium to superfluid helium which cool down the sample to 1.8K and a mechanical heat switch connect or disconnect the cryocooler and the pot. In order to provide the sufficient time for the test, the cooling system is designed to keep the sample at 1.8K for 300 seconds.

1. Introduction

The insulation material is the key elements of the superconducting magnets which is characterized by high mechanical strength in order to withstand the Lorenz forces occurring in the powered magnets, high capacity to transfer heat from the magnet coil to the magnet coolant and high electrical strength. The insulation character prevents electric discharge when the superconducting magnet quenches or discharges [1]. The research on the superconducting magnets insulation should be tested for the cryogenic environment. Several similar tests have been published finding the improvement in the electrical breakdown strength along with the dropping of the temperature. In a system designed by Husain et al [2]. Measurements were taken in the liquid nitrogen for the electrical breakdown voltage of solid film insulation material. They designed different kinds of electrodes to test the sample and found the type of electrodes also affected the breakdown voltage. Zhang Jinxing [3] and co-workers designed a system that reaches 80K in vacuum or helium gas in different pressures which also use liquid nitrogen to cool down the sample. So the electrical breakdown strength measuring system for solid dielectric has not been developed below the temperature of liquid nitrogen. The data is limited on electrical breakdown strength for solid material below the temperature of 77K. This hinders the development of the discovery of the relationship between electrical breakdown strength and the temperature.

So in this paper, we designed a cryostat aimed to test the electrical breakdown character of the thin film insulation materials. Using alone the cryocooler doesn't have the ability to reach the temperature of 1.8K. The cryocooler cannot provide sufficient cooling capacity bearing the heat leakage from the shield radiation and structure heat leakage. So we combine liquid helium and GM cryocooler in a cryostat to cool down the sample to the temperature of 1.8K. A 50K radiation shield is set to block the radiation. We also design a heat switch to control the cool down process. The superfluid helium was made by pumping the liquid helium in a copper pot. After the helium pressure and the temperature of sample become stable, the power supply controlled by computer begins to apply high voltage on the positive electrode until the sample breakdowns.

2. Cryostat design

2.1. The refrigeration system

The cooling process of the system consists of two parts. First, cool down the system by the GM cryocooler to the temperature of liquid helium. Second, inject the liquid helium into the superfluid helium pot and pump to the temperature of superfluid helium.

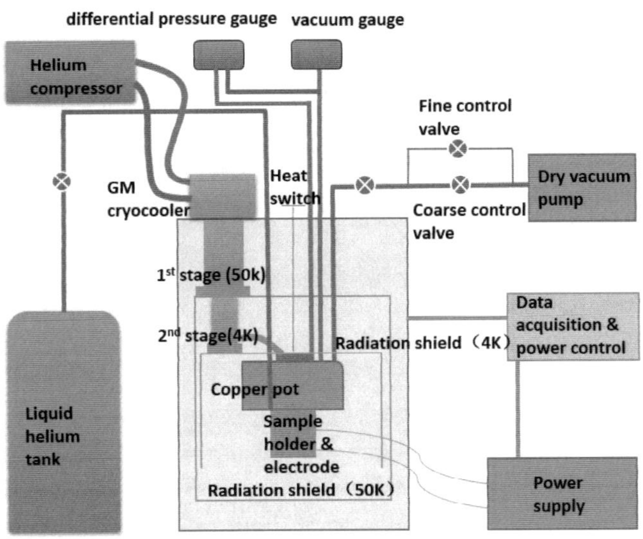

Fig.1 The schematic of the cryostat with the GM cryocooler

Fig.1 shows the schematic of the cryostat with the GM cryocooler. An ESYCOOL KDE415SA two stages GM cryocooler is used to provide refrigeration at 4K. The GM cryocooler specification is: 35W@50K on the first stage and 1.5W@4.2K on the second stage. The cryocooler consists of a helium compressor, flexible connection lines and a cool head. The cool head is made up of two stages. Two radiation shields connect to the first and the second stage separately. The thermal connect between the shields and stages is enhanced by 0.25 thick indium washer and the copper flange was applied with Apiezon grease. The temperature of the shields is about 50K and 4K. The shield is wrapped in ten layers of insulation. The radiation shields cut down most radiant heat leakage.

The cryostat incorporates a liquid helium pot (also called superfluid helium pot, 650ml in maximum). The pot is made of copper. By design, the pot is thermally isolated from the rest of the cryostat. It is fixed to the first stage radiation shield by four epoxy resin rods. The liquid helium is provided by a standard liquid helium tank, it transfuses to the pot through the first pipe (10mm diameter and vacuum double-wall). The liquid level in the pot was detected by the differential pressure gauge from two pipes. The pipes (2mm in diameter) connect to the top and bottom of the pot. The pipe on the top is used to detect the helium gas pressure, and the bottom one is used to detect the bottom liquid helium. The radiation on the third pipe makes the liquid helium evaporate, then the liquid helium cannot inject into the third pipe. The pipe (20mm in diameter and 1mm thick) is used to pump the pot. Because of the 'zero gravity character of superfluid helium, the holes on the pot are limited to 2mm in diameter. Four stainless pipes connect to the pot.

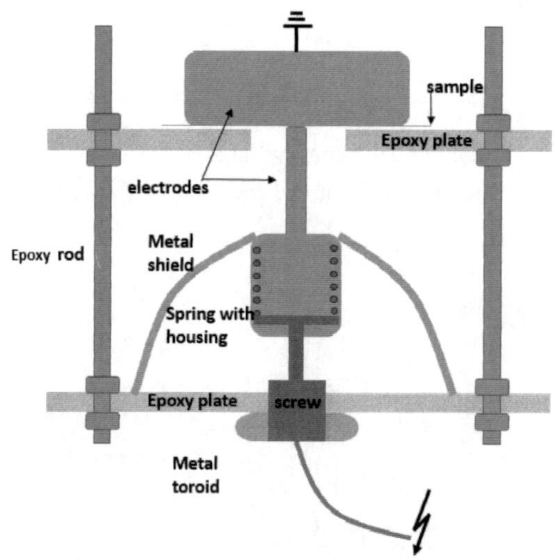

Fig.2 The axial symmetric test electrodes

The pot at the superfluid helium temperature needs to disconnect preventing the heat leak from the second stage. To expedite the cooling process, we designed a simple mechanical heat switch. During the cool down phase, a copper disk is held tightly against the pot by screwing on the room temperature side. Thermal contact between the copper disk and the pot enhanced by a 0.25mm thick indium washer between them. Five flexible braids, each made up of copper wires, connect the disk to the second stage. The switch connects the pot with pressure, allowing the heat efficiently removed from the pot.

2.2. The electrical test system

The high voltage power supply with overload protection provides the high voltage on the sample. The maximum voltage is 100kV and the maximum current is 2mA, so the capacity is enough to break the thin film sample. A protection resistant 100MΩ protects circuits after the sample breaks down, the voltage on the sample was detected by a 1000:1 voltage divider. The signal of the electrical breakdown is an instant fall of the voltage between the voltage divider and ground. The high voltage wire goes into the cryostat by a vacuum feedthrough. Inside the cryostat, the vacuum side of the feedthrough directly connects to the test cell with XLPE. Fig.2 shows the axial symmetric test cell that is installed under the bottom of the superfluid helium pot. 8 tubes and two plates shaped frame are made of glass fiber reinforced epoxy and have a diameter of 10mm. We choose needle to flat electrodes (showed in fig.2 [4]) in a parallel plane configuration according to the IEC standard. The flat electrode integrated with the superfluid helium chamber has enough area to conduct the cool capacity from the bottom of the superfluid helium chamber, and the needle shape reduces the heat from the positive pole wire from the room temperature. The thin film sample to be tested are put between the electrodes. The diameter of the sample should be a little bigger than the electrodes preventing the flash over the surface. The needle electrode is made of brass when tightening the screw going through the thick epoxy plate the spring moves upwards. Thereby compressing the spring and thus exerting this compressive force on the electrodes system through the hemispherical tip of the brass plate. High voltage is provided by an XLPE insulated power cable with a metal toroid above the thick epoxy plate also contribute to smoothen the electrical fields. The insulation method in the cryostat is using the insulation fracture and high vacuum insulation. The high vacuum means 10-4 ~10-5 Pa, the pressure in cryostat is detected by compound pressure and vacuum gauge. The electrical breakdown voltage in a high vacuum is up to 10-7 V/cm, even so all parts close to low potential conductive material in cryostat should smooth the electrical.

3. Heat load

The cryostat provides vacuum environment for the thermal insulation. The superfluid helium pot operated at 1.8K, the first stage radiation shield is 50K and the second radiation shield is 4.2K. The steady-state heat load, whether radiate or conduction, are driven by the temperature difference between the vacuum vessel and the helium tank. In order to realize producing the long durable 1.8k environment, minimizing the heat leak and thus minimize active cooling power requirements and the consuming of the helium.

We calculated the number of the steady-state heat load roughly. The heat load from heat conduction is calculated by the basic Fourier's equation [5]. The heat leak of radiation through the multi-layer insulation could be calculated from the Lockheed equation [6]. The temperature in the equation ($T_H, T_c = (300,50,4.2,1.8)$) changes when calculate different part. The number of the layers is 30 and it roughly equals to the number of layer density ($N_S = \overline{N} = 30$). The hemispherical emittance ($\varepsilon_{RT} = 1.0$) we choose is 1.0 because of the polished metal. The inner surface areas (A=(1.31, 0.6, 1.1×10^{-2}))of the different shields are 1.31, 0.6 and 1.1×10^{-2}. The degradation (F=1.2) is 1.2. The results are shown in table1.

Table 1. Heat loads to the first, second stages of GM cryocooler and superfluid helium

Item	First stage (W)	Second stage (W)	Helium pot (W)
Radiation	1.36	7.48×10^{-2}	9.07×10^{-4}
Support	1.53	2.65×10^{-3}	1.64×10^{-3}
Total	2.89	7.75×10^{-2}	2.55×10^{-3}

$$q = \frac{C_s(\overline{N})^{2.56}T_m}{N_s+1}(T_m - T_c) + \frac{C_r\varepsilon_{RT}}{N_s}(T_H^{4.67} - T_c^{4.67}) \quad (1)$$

Where

$$C_r = 5.39\times10^{-10}$$

$$C_s = 8.95\times10^{-8}$$

$$\overline{N} = \text{Radiation shield layer density, layers/cm}$$

$$N_S = \text{Number of radiation shields}$$

$$q = \text{Heat flux, W/M}^2$$

$$T_c = \text{Cold boundary temperature, K}$$

$$T_H = \text{Warm boundary temperature, K}$$

$$T_m = \text{Mean insulation temperature, K}$$

$$\varepsilon_{RT} = \text{Room temperature (300K) total hemispherical DAM emittance}$$

$$Q = FAq \quad (2)$$

Where

$$A = \text{Inner surface area}$$

$$F = \text{Degradatai on factor}$$

4. Conclusion

The 1.8K breakdown voltage measurement system with cryocooler is designed. Innovation point of this design is combining the liquid helium and GM cryocooler, which have the ability to reach the temperature of 1.8K theoretically. A copper pot connecting to the second stage by a heat switch cooled by the cryocooler. The heat switch prevents the heat leak from the cryocooler. The sample holder with high voltage poles is located under the copper pot, the liquid helium in the pot is pumped to cool down the sample by heat conduction. The maximum voltage is 60kV and the cooling time is 10 minutes. The test system is under constructing and the result curve will be published in the future.

Acknowledgments
This work was supported by the National Natural Science Foundation of China (Grant Nos. 51522705, 51577185, 51427806, 51408586), the National Magnetic Confinement Fusion Science Program (Grant No. 2015GB121001) and The Found of CAS, (Grand No. QYZDB-SSW-JSC042).

References

[1] Evans D 2001 *Physica C* **354** 136
[2] Husain E, Mohsain M M, Masood A, Zuberi M U and Alam M S 2002 *IEEE Trans. Dielectr. Electr. Insul* **9** 932
[3] Jinxing Z 2014 *J. Fusion Energ* **33** 594
[4] Runde M, Magnusson N and Lillevik O *IEEE Trans. Dielectr. Electr. Insul* **13** 1371
[5] Incropera F P, DeWitt D P, Bergman T L, Lavine A S 2007 *Fundamentals of Heat and Mass Transfer* 6th edition (Jefferson City: John Wiley & Sons) p 59
[6] Nast T C, Frank D J and Feller J 2014 *Cryogenics* **64** 105-1

IOP Conf. Series: Journal of Physics: Conf. Series **897** (2017) 012015 doi:10.1088/1742-6596/897/1/012015

Commissioning of the JT-60SA helium refrigerator

Koji Kamiya[1], Kyohei Natsume[1], Kiichi Ohtsu[1], Makoto Oishi[1], Atsushi Honda[1], Yoshitoshi Kashiwa[1], Kaname Kizu[1], Yoshihiko Koide[1], Christine Hoa[2], Frederic Michel[2], Pascal Roussel[2], Valerie Lamaison[2], Francois Bonne[2], Enrico Dipietro[3], Antonino Cardella[3], Manfred Wanner[3], Jerome Legrand[4], Vincent Pudys[4], Baptiste Langevin[5]

[1] Quantum and Radiology Science and Technology, 801-1 Mukoyama, Naka, Ibaraki 311-0193, Japan
[2] CEA, 91191 Gif-sur-Yvette cedex, France
[3] Fusion for Energy, c/ Josep Pla, nº 2, Torres Diagonal Litoral, Edificio B3, 08019 Barcelona, Spain
[4] Air Liquide Advanced Technology, 2, rue Clémencière, BP 15, 38360 Sassenage, France
[5] Engie Ineo, 1 place des degrès, 92 059 Paris la Défense Cedex, France

kamiya.koji@qst.go.jp

Abstract. The JT-60SA project will use superconducting magnets to confine the plasma and achieve a plasma current with a typical flat top duration of 100 second in purely inductive mode. The helium refrigerator has an equivalent cooling power of 9 kW at 4.5 K providing 3.7 K, 4.5 K, 50 K and 80 K for the diverter cryopump, the superconducting magnets, the HTS current leads, and the thermal shields, respectively. This paper summarizes the JT-60SA helium refrigerator commissioning activities aiming at successful operation of heat load smoothing technology to manage the 12 kW heat pulses by 9 kW cooling power using a 7000 liter liquid helium.

1. Introduction

The JT-60U is being upgraded to a full-superconducting tokamak referred as the JT-60 Super Advanced (JT-60SA) as one of the JA-EU broader approach projects [1]. The JT-60SA project will use superconducting magnets to confine the plasma and achieve a plasma current with a typical flat top duration of 100 second in purely inductive mode. The helium refrigerator provides 4.4 K, 50 K, and 80 K for the superconducting magnets, the HTS current leads, and the thermal shields(TS), respectively. In addition 3.7 K helium is provided for the divertor cryopumps. The refrigerator consists of warm compressors, a refrigeration cold box, an auxiliary cold box and LN2 and GHe storages.

The refrigerator was manufactured and installed in Naka Fusion institute by Air Liquid Advanced Technology, France in September 2015 [2]. The commissioning started in October 2015 and has been finished in October 2016. During the commissioning the followings operations were tested; automatic cool-down and warm-up, automatic purification, the total cooling capacity, stable operation in the

Short Term Stand-by (STS) and baking (BAK) mode, nominal pulse modes and quench sequence, regeneration sequence of the divertor cryopanels. In addition, automatic transition tests between all defined stationary modes, such as 80 K stand-by mode, STS mode, BAK mode, TF nominal mode, Magnet nominal mode, etc. have been also checked and validated during the commissioning.

2. JT-60SA helium refrigerator

The JT-60SA refrigerator is the world largest class helium refrigerator for a fusion device with an equivalent cooling power of 9 kW at 4.5 K[3,4]. One of the feature of the refrigerator is the capability of the large transient heat pulses by the heat load smoothing technology utilizing a 7000 liter liquid helium damper. The helium bath absorbs the pulse heat by a temporary temperature increase and subsequent vaporization of helium liquid [5]. After the pulse, the refrigerator starts liquefying to recover the liquid level to the original for the next pulse. The JT-60SA refrigerator also has the capability of supplying 50 K for HTS-CL, and the temperature-controllable automatic cool down and warm-up, which is the first challenge by a liquid nitrogen pre-cooled refrigerator.

Figure 1 shows a whole view of the JT-60SA helium refrigerator. The refrigerator will use 1250 m3 gaseous helium stored in 5 helium storage vessels. There is an extra helium gaseous tank dedicated for magnet quench, which is specially designed to avoid local excess cooling of the tank by large amount of cold quench helium gas. In the quench tank, a perforated nozzle is inserted to spray the gas homogenously. For pre-cooling of the refrigerator in the commissioning, 65000 liter liquid nitrogen tank is used. Because even STS mode consumes over 30000 liter nitrogen a day, which is more than half of the tank capacity, an additional 100000-liter liquid nitrogen tank is planned to install next to the existing 65000 liter tank for more stable and reliable tokamak operation.

Figure 1. A whole view of the JT-60SA refrigerator

Figure 2. A layout and pictures of the refrigerator and its components. A top figure is a whole layout with close-up of the cold boxes and compressors. Corresponding pictures for components are given below the layout drawing.

IOP Conf. Series: Journal of Physics: Conf. Series **897** (2017) 012015 doi:10.1088/1742-6596/897/1/012015

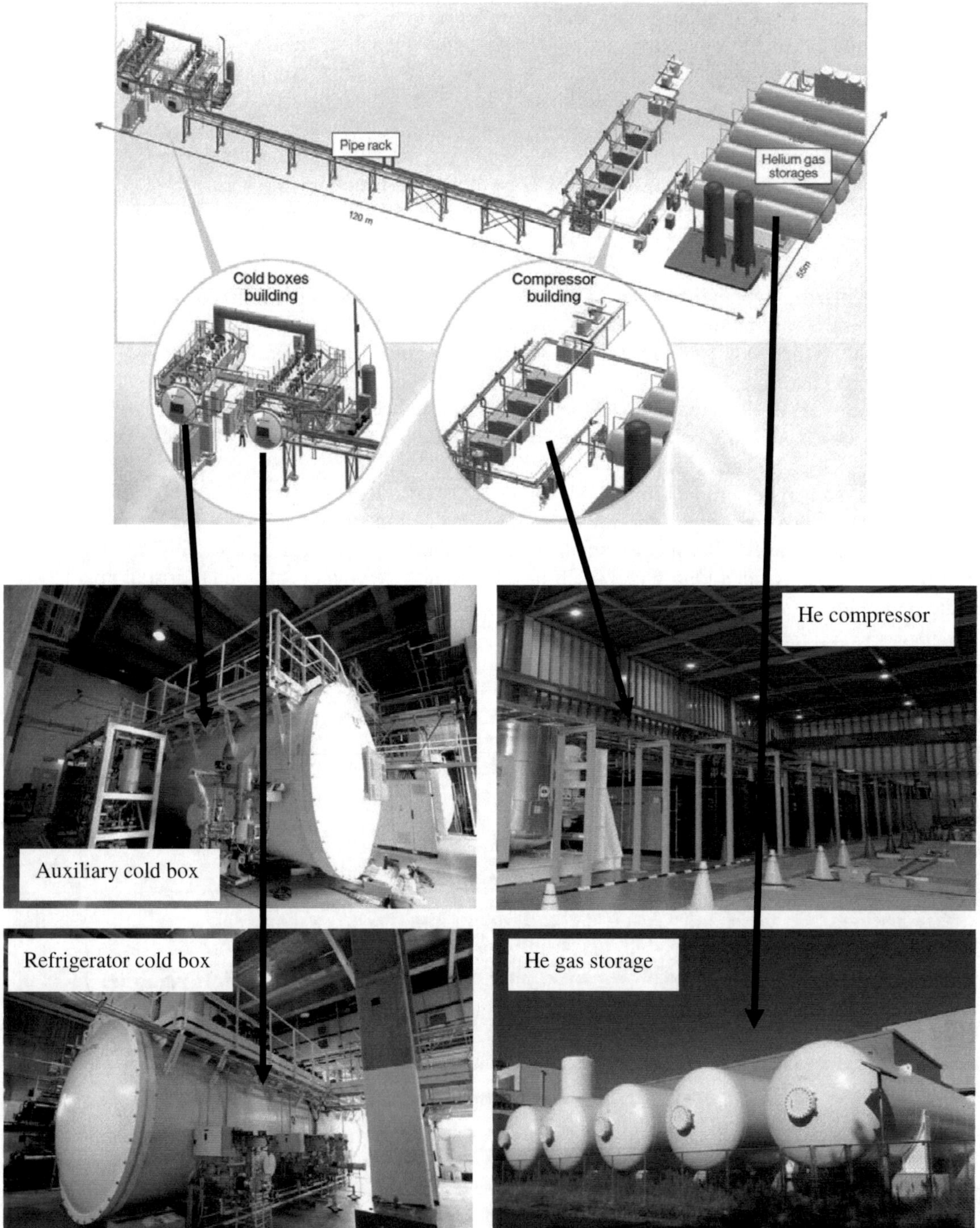

Figure 2. Layout (top) and pictures of the JT-60SA refrigerator components

Table 1 shows the main cryogenic user the JT-60SA refrigerator provides the cooling powers. The 650 ton superconducting magnet consists of the Toroidal Field Coil(TFC), the Central Solenoid(CS), and the Equilibrium Field Coil(EFC) at 4.5 K is the dominant mass of the cryogenic user. The coil mass consists of its Winding Pack(WP) and the structure(STR). The second dominant mass is the thermal shield of 100 ton at 80 K. Table 2 shows requirements of the JT-60SA refrigerator in 3 different operation modes, Plasma Operation State(POS), Short Term Standby(STS), Baking mode (BAK). The POS corresponds to the pulse heat operation, which is most demanding operation. The STS mode is operated during the night every day after daily operation.

Table 1. Main cryogenic users mass

Cryogenic user	TFC (STR/WP)	CS (STR/WP)	EFC (STR/WP)	Total magnet	Cryopump panels/baffles	Thermal shield
Weight(ton)	280/100	33/60	64/114	650	1.3	100
Material	Stainless steel/Cu	Stainless steel/Cu	Stainless steel/Cu	Stainless steel/Cu	Stainless steel	Stainless steel

Table 2. shows requirements of the JT60-SA refrigerator in 3 different operation modes.

Cryogenic user	Temperature	POS	STS	BAK
Thermal shields, cryopump baffles	80 K (140 K for BAK)	42000 W	41300 W	135000 W
HTS current leads	50 K	25 g/s	13 g/s	0 g/s
TFC and structure	4.4 K (20 K for BAK)	1794 W	1270 W	2260 W
CS and EFC	4.4 K (20 K for BAK)	1850 W	600 W	1130 W
Cryopump panel	3.7 K	84 W	Regeneration	0 W

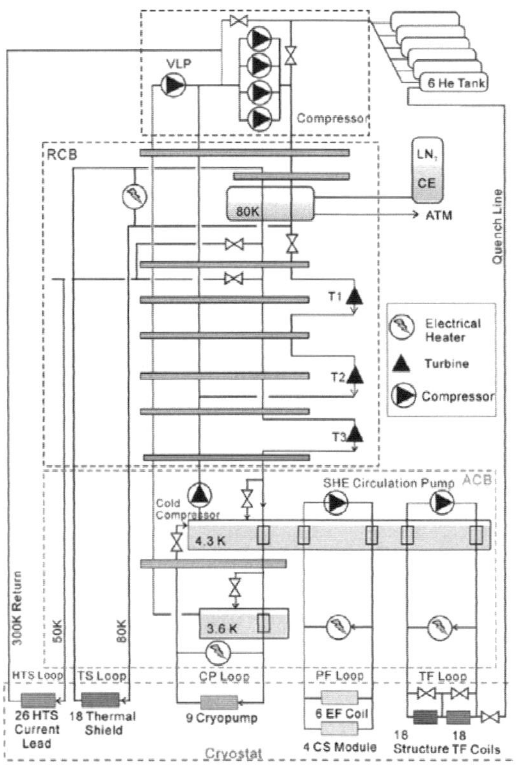

Figure 3. Process diagram of the JT-60SA refrigerator

During the night STS, the cryopump panel planned to be regenerated for 7 hours. Once a month, the plasma vacuum vessel of the JT-60SA plans to be in BAK mode at 200 degree in Celsius for one week. During the BAK, the thermal shield outlet temperature goes up to 140 K.

Figure 3 shows the process diagram of the JT-60SA refrigerator in order to achieve the requirement in table 2. Stored gaseous helium, first, is transferred to the 4 KAESER compressor units to be pressurized up to 1.5 MPa. The pressurized helium is pre-cooled with liquid nitrogen, and expanded by 3 turbines exchanging the heat by the heat exchangers. The helium is supplied to the 4.4 K damper with volume of 7000 liters and 3.7 K damper. The 4.4 K damper pressure and temperature is controlled by the cold compressor located at top left of the damper. This cold compressor also plays very important role for control of the pulse heat operation. The JT-60SA refrigerator has 2 circulators for TFC/Structures and PF coils as shown in Figure 3. In order to obtain 3.7 K, the refrigerator has the VLP(Very Low Pressure) pump at room temperature.

3. Tests in commissioning

In the commissioning of the JT-60SA refrigerator, following modes were tested, automatic cool-down and warm-up with temperature controlling, automatic purification, total 6.6 kW cooling capacity, stable operation in STS and BAK, pulse tests, quench sequence. In this paper, result of cool-down and pulse test are briefly reported. Other tests will be presented elsewhere.

3.1. Cool-down

One of the conditions of the temperature control of cool-down is that during the cool down the temperature of the magnets is maintained 20K - 50K lower than that of the thermal shields down to 80 K, in order to trap all impurities on the colder surfaces and maintain the emissivity of the polished TS surface. The second condition is that the temperature difference between inlet and outlet temperature (of TS and Coil) is always smaller than 40 K to avoid excess thermal stresses. In cool-down, heats were applied to the coil and the TS to simulate the Tokamak to some extent. Results of cool down test is shown in Figure 4. In the cool-down, the coils are firstly cooled followed by the thermal shields active cooling as is designed. It is found the temperature of the coil is always 20 K below the temperature of the thermal shields. It is also found that the temperature difference between outlet and inlet temperature of the thermal shields are always below 40 K. It was concluded that cool down test was successfully performed by defined temperature control.

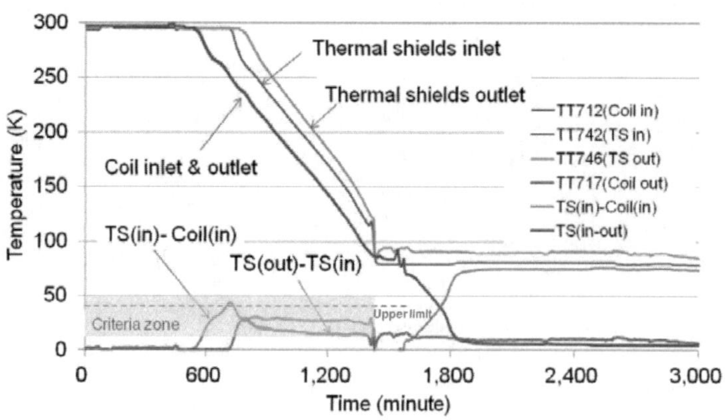

Figure 4. Temperature evolutions of the TS and coil in cool down

3.2. Pulse test

JT-60SA shall be capable of investigating several design scenarios. There are mainly 2 kinds of repetition time; 1800 second for 60 second pulse and 3000 second for 100 second pulse. Figure 5

IOP Conf. Series: Journal of Physics: Conf. Series **897** (2017) 012015 doi:10.1088/1742-6596/897/1/012015

shows schematic of the pulse heat control. In this paper, the result of 60 second pulse is briefly presented. In the pulse tests, a decayed heat profile after travelling all the way in the transfer line is calculated to input to 2 loops for TF and PF in the ACB as shown in top in Figure 5.

The pulse heat load for each loop consists of the conductor AC loss, conductor joint heat, nuclear heating, the structure eddy current in addition to the steady heat load such as the conduction and the radiation heat loss [6]. When the pulse arrives the damper of the refrigerator, the pulse heat will increase the damper temperature and the damper pressure increases accordingly. Subsequently, the cold compressor slows down its speed to keep the energy in the damper. This is mainly to keep the pressure in the low pressure line as stable operation of the compressors as possible. The increased damper temperature expands the liquid by thermal expansion by several percent. After passing the pulse, the cold compressor start speed up its speed again to go back to the initial conditions. In the right bottom of figure 5, it is found the damper pressure and the temperature with liquid level successfully recovered its initial position after the pulse as is required in the criteria.

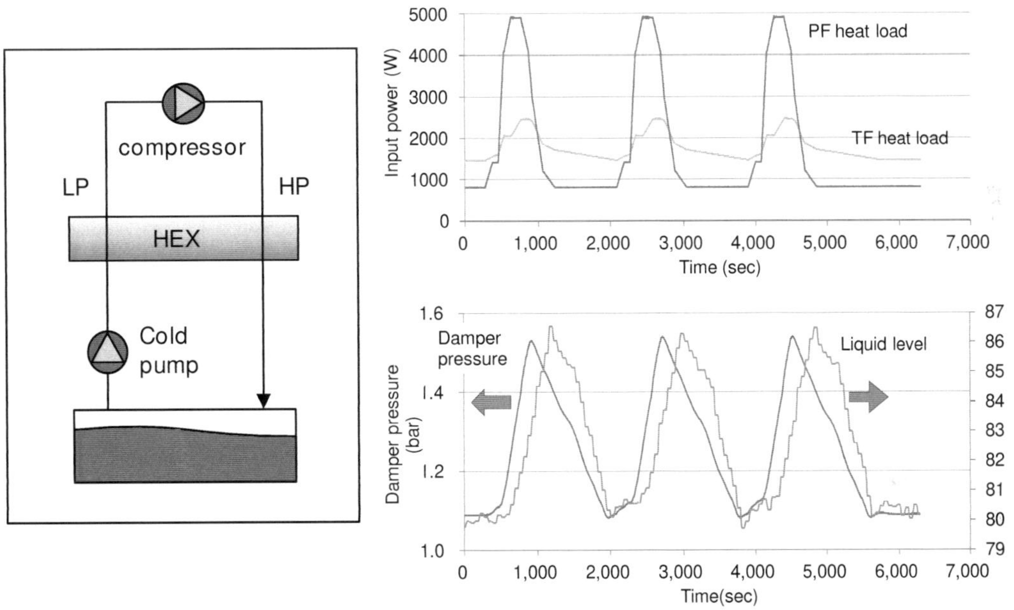

Figure 5. Schematic of pulse heat control(left), input heat pulses(right top), and test results(right bottom)

In the commissioning, auto purification test, 6.5 kW capacity test, quench tests, and cryopump regeneration test are also performed to confirm the required specification of the JT-60SA helium refrigerator.

4. Utility consumption

For the commissioning, 13500 Nm^3 (2.4 tons) gaseous helium was first supplied to the refrigerator. Total consumption of the liquid nitrogen and electrical power during the commissioning reaches 1300 m^3 and 6700 MWh, respectively. The flow rate of the cooling water is 200 m^3/h and 12 m^3/h for the cycle compressors and turbines(pumps, circulators), respectively. Current helium inventory is 9500 Nm^3 (1.7 tons), which is 70% of initial inventory.

Table 3 shows resulted consumption of electrical power and liquid nitrogen in different mode of the refrigerator. It is found these result are in good agreement with the design values. In the cool-down of the commissioning, the refrigerator consumes rather small amount of nitrogen because no heat loads from coils and thermal shields are simulated by heaters. This should go up to around 1200 liter/hour with Tokamak, which is the same as the STS. In the STS and the BAK in table 3, the estimated heater powers were applied to the coils and the thermal shields to see the stability of the refrigerator. In the BAK test, 10 trucks(capacity of 8000 liters) continuously deliver the LN_2 in a day.

Table 3. Utility consumption in each mode

Utility	unit	Cool-down	STS	BAK	POS	Warm-up
LN_2	liter/hour	400	1200	3400	1400	400
Electricity	MW	1.6	2.1	2.0	2.6	1.9

5. Conclusions

All the tests planned in the commissioning of the JT-60SA helium refrigerator was successfully finished as follows:

- Automatic cool down was performed under the temperature conditions of the coil and the thermal shields.
- Pulse heat was repeatedly applied at 4.4 K loops. It was confirmed increased temperature, liquid level, vapor pressure in the damper were recovered to the initial level within 30 minutes.
- Other tests such as the 6.5 kW cooling capacity test, STS/BAK tests with stability criteria for 36 hours, quench sequence test, were all performed successfully.

It can be concluded that heat smooth technology was successfully demonstrated to manage the 12 kW heat pulse with only 9 kW cooling power using 7000-liter liquid helium damper.

References

[1] Ishida S, Barabaschi P, Kamada Y and the JT-60SA Team, Status and prospect of the JT- 60SA project *Fusion Engineering and Design* pp 2070-2079 (2010)

[2] Hoa Christine et al., Installation and pre-commissioning of the cryogenic system of JT-60SA tokamak *Proceedings of International Cryogenic Engineering Conference* to be published.

[3] Michel F et al., Cryogenic Requirements for the JT-60SA Tokamak *Advances in Cryogenic Engineering* Volume 57, pp. 78-85 (2012)

[4] V Lamaison, J. Beauvisage, P. Fejoz, S. Girard, Gonvalves, P. Gonde, V. Heloin, F. Michel, C. Hoa, K. Kamiya, P. Roussel, J-C. Vallet, M. Wanner and K. Yoshida, Conceptual Design of the JT-60SA Cryogenic System *AIP Conf. Proc.* 1573, p337 (2014)

[5] Hoa C et al., Experimental and numerical investigations for the operation of large scale helium supercritical loops subjected to pulsed heat loads in tokamaks, *Physic Procedia* Volume 67 pp. 54–59 (2015)

[6] V. LAMAISON, D. HITZ, C. HOA, F. MICHEL, P. REYNAUD, P. ROUSSEL, M. WANNER, Determination of heat loads at the interface of the JT-60SA cryogenic system. ICEA23-ICMC2010, July pp.19-23 (2010)

A Mechanical Thermal Switch for Conduction-cooled Cryogenic System

Monan Li[1,2], Lifeng Li[1,2], Dong Xu[1]

[1]Key Laboratory of Cryogenics, Technical Institute of Physics and Chemistry, Chinese Academy of Sciences, Beijing, China
[2]University of Chinese Academy of Sciences, Beijing, China

Xudong@mail.ipc.ac.cn

Abstract. A mechanical thermal switch is designed and fabricated to shorten the cool-down time of a conduction-cooled cryogenic system in which several copper heat loads are selected as experimental object. It is the first try of mechanical thermal switch for this application. Compared with the behavior of gas-gap thermal switch, the mechanical thermal switch is easy to manufacture, able to turn off at any temperature. The heat leak is little when it turns off. The drawback is that mechanical thermal switch need people to operate. The result shows that gas-gap thermal switch reduces the cooling time of our experimental system from 41.15 hours to 35.16 hours, while mechanical thermal switch reduces the time to 30.9 hours. In the situations for which frequent switches not needed, mechanical thermal switch is a competitive choice.

1. Introduction

The conduction-cooled cryogenic system, using a compact cryocooler instead of cryogenic liquid, is getting popular due to its compactness, easy and safe handling and low running cost. A typical conduction-cooled cryogenic system includes the object to be cooled, thermal radiation shield, GM cryocooler or Pulse-tube cryocooler, vacuum vessel, and other accessories. Thermal radiation shield is connected with the first stage of cryocooler and cooled by its cooling power, whereas the lowest temperatures of the closed-cycle refrigerator are reached at the second stage where experiments are attached. In commercial two-stage closed-cycle cryocoolers, the first stage is always considerably more powerful than the second stage over a wide temperature range. When an object with a large heat capacity has to be cooled by the second stage, the temperature of the first stage approaches its final temperature quickly, whereas it can take many hours for the second stage to cool the exprimental object with its high heat reservoir to its lowest temperature. In order to reduce the initial cool-down time of conduction-cooled cryogenic system, a thermal switch can be installed between the first and second stage of a cryocooler, or between the first stage and the object. At on-state, the thermal switch is a very efficient heat transfer device, therefore, the large cooling capacity of the first stage can be used to cool the object to temperature as low as possible for precooling. Once the combined stages reach the final temperature of the first stage, the thermal switch has to be interrupted and the second stage can continue to cool to its final temperature.

There have been a number of previous works to shorten the cool-down time of conduction-cooled cryogenic system:

1) Gas-gap thermal switch (GGTS). The presence/absence of the gas in the small gap enables/disables a thermal contact. It is divided into active and passive GGTS depending on if there are adsorbers or valves connected to gas reservoirs.

1.1) Passive GGTS. At cryogenic temperature, the thermal isolation (off-state) can be achieved without any external actuation, because the gas is frosted and the corresponding vapor pressure is significantly decreased.

Bywaters [1] firstly came up with a gas-gap thermal switch in 1973. A gas-gap thermal switch filled with helium was designed, fabricated and tested. The switch resistance ratio, Roff/Ron= 512 when the helium gas-gap was 0.254mm. In 2000, Ho-Myung Chang [2] fabricated a new configuration for the switch and tested the new one and the conventional structure switch in a G-M cooler cooled superconducting magnet system. The experiment showed that the newly developed gas-gap switch achieved a superior thermal performance. For both switches, the overall cool down time was shorten (shown in Fig.1). In 2012, Wang Q.L. [3] checked the performance of a gas-gap thermal switch filled with air using a superconducting magnet whose weight was 41.5kg as background device. The experiment showed that the cool-down time was shortened from 25 hours to 20 hours.

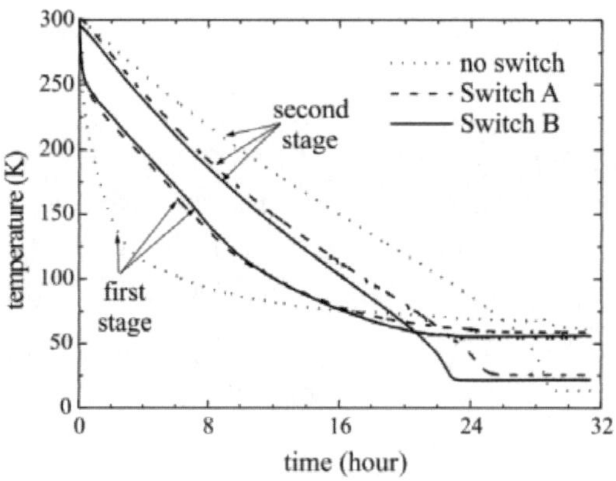

Figure 1. Performance of two gas-gap thermal switches. [2]

1.2) Active GGTS. The pressure of a gas in a small gap is controlled by use of adsorbers or valves connected to gas reservoirs. J Barreto[4] developed and built a prototype of a heat switch to be implemented at the Inter University Accelerator Centre according to thermal and mechanical requirements of the magnet cryostat coupled with a 1.5 W @ 4.2 K Gifford-McMahon CCR.

2) Detachable thermal switch. In 2006, a group in Korea[5] uses a thermosiphon between the first and second stages of a cryocooler in a conduction-cooled superconducting magnet. It can be detached mechanically from the magnet. This year, D.Xu in our laboratory used a mechanical-thermal switch worked as a novel pre-cooling system for a conduction-cooled test bed.

3) Thermal shunt. Kurt Uhlig [6] tested an efficient and easy-to-build gas circuit acting as a thermal shunt between the two stages of a Gifford–McMahon cooler during the cool-down period of an experiment.

Mechanical thermal switch is a kind of detachable thermal switch. Here we designed a mechanical thermal switch and a gas-gap thermal switch to shorten the cool-down time of a conduction-cooled cryogenic system with copper blocks acting as heat load, and compared their performances.

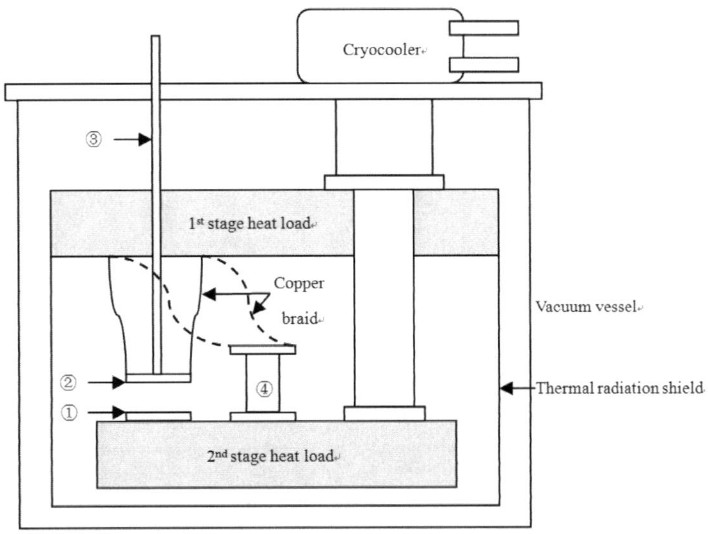

Figure 2. Schematic diagram of experimental setup: ①② copper blocks of mechanical thermal switch; ③ switch rod; ④ gas-gap thermal switch.

2. Experimental apparatus

2.1. Cryocooler and heat loads

A two-stage GM cryocooler (RDK415D, Sumitomo Heavy Industries, Ltd., Japan) with cooling power around 1.5 W at 4.2 K is used. The copper block and thermal radiation shield connected with the first stage of the cryocooler compose the 1st stage heat load which weighs 56 kg in total. The copper block connected with the second stage of the cryocooler which weights 42 kg acts as the 2nd stage heat load. Between the 1st and 2nd stage heat loads and cold flanges, a thin indium foil and contact grease are applied to improve thermal contact. Both heat loads are covered with 30 layers of aluminium-foil paper to reduce the radiation heat. Temperature measurements of 1st and 2nd stage heat loads (T1 and T2) are performed with two standard Rhodium–Iron resistance thermometers plugged in the holes of the two copper blocks, and the lead wires are thermally sunk to the thermal anchor. The conductivity of copper is better than actual experimental object, so a piece of teflon is placed between the 2nd stage heat load and cold flange to add a thermal resistance. The size of the teflon is the same as 2nd stage cold flange, and the thinkness is 0.5mm.

2.2. Mechanical thermal switch

The mechanical thermal switch is thermally connected to the 1st stage heat load through 12 copper braids. Every two copper braids are welded together at two ends with copper terminal in order to reduce the thermal resistance of braids by increasing the cross section area. Copper block ① is tightly connected with the 2nd stage heat load through indium foil and contact grease. On the top surface of copper block ①, indium foil and contact grease are also used to optimize thermal contact with copper block ② when the mechanical thermal switch is on. Copper block ② could be moved up and down to keep thermal contact or not with block by spinning the end of switch rod at room temperature. Thus, the mechanical thermal switch could be turned on during the pre-cooling process and off during the test process.

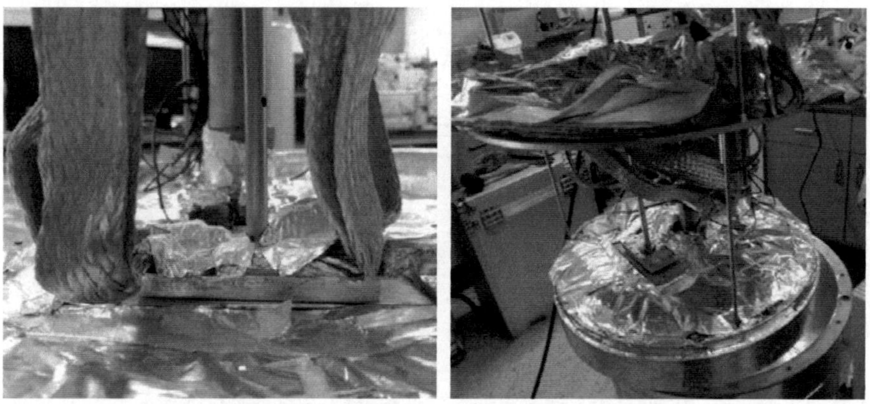

(a) (b)

Figure 3. Mechanical thermal switch.
(a) mechanical thermal switch with copper braids;
(b) mechanical thermal switch without copper braids in test-bed.

2.3. Gas-gap thermal switch

The structure of gas-gap thermal switch is shown in Fig. 4. It is actually a closed container, composed of two caps with several straight and parallel fins which are made of copper. The 1 mm thick fins are inserted each other and form gas-gap space. The outer shell is made of stainless steel and its thickness is 0.3 mm. The outer diameter of the switch is 37.6 mm and the total height is about 64 mm. The gap between adjacent walls is 1 mm, and filled with nitrogen.

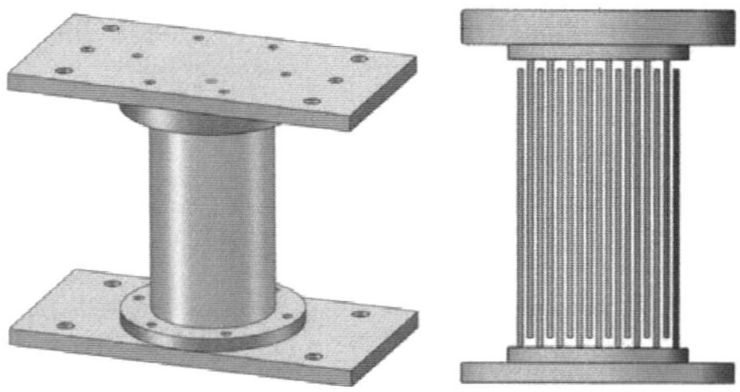

(a) (b)

Figure 4. Gas-gap thermal switch.
(a) appearance; (b) internal structure without outer shell.

3. Experimental procedure

3.1. No thermal switch

In this experiment, both thermal switches are installed in the system, while no copper braids connect them with 1st stage heat load. The thermal switches act as a part of 2nd stage heat load here. After the inside of vacuum vessel is pumped to vacuum ($\leqslant 10^{-4}$ Pa at room temperature), GM cryocooler is turned on to cool the 2nd stage heat load to 50K.

3.2. Mechanical thermal switch

After the temperature of the experiment system back to room temperature, the mechanical thermal switch is connected with the 1st stage heat load through aforementioned 12 copper braids. Contact grease is used to reduce thermal contact resistance between both ends of copper braids and 1st stage heat load or the upper copper block of mechanical thermal switch. Then the end of switch rod is spinned to guarantee tight junction of two copper blocks. Repeat the operation in the first experiment to cool the 2nd stage heat load to 50K.

3.3. No thermal switch

The system is warmed to room temperature, and move one end of the copper braids from mechanical thermal switch to gas-gap thermal switch. The thermal switch is purified with 99.999% nitrogen gas and charged to 150 kPa. Then the operation in the first experiment is repeated to cool the 2nd stage heat load.

In addition, the experimental setup comprises measurement devices and a data acquisition system. The control electronics for the measurement are driven by a data collection computer via the IEEE-488 General Purpose Interface Bus (GPIB). Based on Labview Software, a custom designed data acquisition program is developed that automates the data acquisition process.

4. Results and discussion

Figure 5 shows the cool-down curves of 2nd stage heat load: (1) no thermal switch; (2) connecting 1st stage heat load and mechanical thermal switch with same copper braids; (3) connecting 1st stage heat load and gas-gap thermal switch with 12 copper braids. In the case of no thermal switch, it takes 41.15 hours to cool down 2nd stage heat load to 50K. With gas-gap thermal switch, it takes 35.16 hours to cool down to 50K. While the total cool down time is 30.9 hours using mechanical thermal switch.

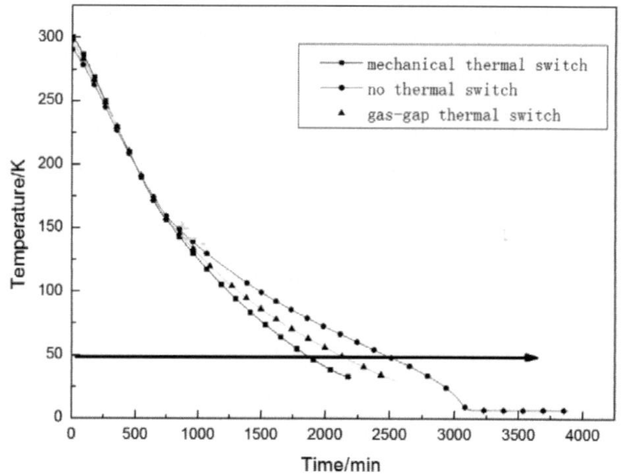

Figure 5. Cool-down curves of 2nd stage heat load

Table 1. Cool-down time of three experiments

	Initial temperature	Target Temperature	Cool-down time
No thermal switch	290 K	50	41.15 h
Gas-gap thermal switch	298 K	50	35.16 h
Mechanical thermal switch	300 K	50	30.9 h

The mechanical thermal switch can significantly shorten the cool-down time from 41.15 h to 30.9 h by making use of the large cooling power of the first stage. Comparing two switches, mechanical thermal switch performs better. In order to make full use of the cooling power of the first stage, we hope the switch can turn off at the final temperature of the first stage. The switching temperature of a passive gas-gap thermal switch depends on the property of the filling gas, while the mechanical thermal switch can turn off at any temperature we want, so that the large cooling power of 1st stage can be used to cool the 2nd stage heat load to temperature as low as possible. In addition, mechanical thermal switch is much easier to fabricate. But mechanical thermal switch need people to turn it off and on, which limits its application. While in some applications, like conduction cooled superconducting magnet, where it does not need to switch frequently, mechanical thermal switch is more flexible. Meanwhile, the heat leak is little at off state. Later, we will test the performance of mechanical thermal switch in a real superconducting magnet.

5. Conclusions

In order to use mechanical thermal switch in some specific applications, such as superconducting magnet, we test its performance in a system whose heat loads consist of several red copper blocks and compare with the performance of a gas-gap thermal switch. Three experiments are carried out: no thermal switch, with gas-gap thermal switch, with mechanical thermal switch. The cool-down time from room temperature to 50K is 41.15 h, 35.16 h and 30.9 h respectively. The result shows mechanical thermal switch can significantly shorten the cool-down time of the conduction-cooled cryogenic system. In some applications such as superconducting magnet, where the system do not need to be cooled down frequently, mechanical thermal switch is a good choice to reduce the cool-down time, because it can be turned on and off at any temperature and is easy to fabricate.

Acknowledgements
The authors appreciate the financial support from the National Natural Science Foundation of China (Grant No.: 51677186).

References

[1] R.P. Bywaters, R.A. Griffin 1973 *Cryogenics* **13** 344-349
[2] H.-M. Chang, H.-J. Kim 2000 *Cryogenics* **40** 769-777
[3] B. Zhao, Q. Wang, L. Li, H. Liu, S. Chen, Y. Dai, Y. Lei, H. Wang, Z. Ni 2012 *IEEE Transactions on Appl. Supercon.* **22** 4700904-4700904
[4] J. Barreto, P.B. de Sousa, D. Martins, S. Kar, G. Bonfait, I. Catarino 2015 *Mater. Sci. & Eng.* **101** 012144
[5] S. Jeong, Y. Kim, C. Noh, S. Kim, H. Jin 2006 *Cryogenics* **46** 705-710
[6] K. Uhlig 2002 *Cryogenics* **42** 67-69

The influence of winding direction of two-layer HTS DC cable on the critical current

V S Vyatkin[1], K Kashiwagi[2], Y V Ivanov[1], E S Otabe[2] and S Yamaguchi[1]
[1]Chubu University, Japan, [2]Kyushu Institute of Technology, Japan

(vyatkin@isc.chubu.ac.jp)

Abstract. The design of twist pitch and direction of winding in multilayer HTS coaxial cable is important. For HTS AC transmitting cables, the main condition of twist pitch is the balance of inductances of each layer for providing the current balance between layers. In this work, the finite element method analysis for the coaxial cables with both same and opposite directions winding is used to calculate magnetic field distribution, and critical current of the cable is estimated. It was found that the critical current of the cable with same direction winding is about 10 percent higher than that in the case of the cable with the opposite direction winding.

1. Introduction

Optimization of coaxial cable winding is necessary for HTS cable designing. For example, the force-free configuration of DC coaxial cable [1, 2] enhances the current-carrying capacity. However, the inductance of different layers is not the same. The same inductance of layers is important to provide the same transport current in each layer for AC cables. In conventional coaxial two-layer cable, each layer has different winding directions [3]. However, it is considered that the perpendicular (*c*-axis) component of magnetic field to the superconducting tape is large in the opposite winding direction, resulting in low critical current property due to the large perpendicular component of magnetic field. Therefore, same winding direction in two-layer cable is considered to be suitable to reduce the perpendicular component of magnetic field and critical current of the cable will be improved.

In the present work, the two-layer HTS cable with same and opposite winding directions in each layer for providing the same layer inductance are used, and the magnetic field distribution around the cable is calculated by finite element method (FEM). The advantageous and disadvantageous of same and opposite directions of winding from the view point of the current carrying capacity of the cable are discussed. The same direction winding provides also longitudinal magnetic field effect. However, it is possible to obtain increasing of the critical current without using the longitudinal magnetic field effect, when critical current of HTS tape higher in the parallel magnetic field. Therefore, this work uses only perpendicular magnetic field dependence of critical current, and the longitudinal magnetic field effect is omitted.

2. Calculation model

The model for calculating the magnetic field distribution and current carrying capacity was the coaxial two-layer cable which inner layer radius was 12 mm and outer layer radius was 13 mm. The specification of parameters of the cables given in the Table 1.

Table 1. The specification of the samples for calculation.

Sample of HTS cable	Number of HTS tapes of inner/outer wire	Diameter of inner /outer layer, [mm]	Twist pitch of inner/outer layer, [mm]
Same direction winding	12/12	12/13	100/100
Opposite direction winding	12/12	12/13	100/100

The calculation performed with program JMAG [4] for finite element method analysis. The layers are not connected electrically and winded in the same or opposite directions. The cross section is shown in figure 1. The twist pitch of the superconducting tape layer is 100 mm, and *z*-axis is the longitudinal direction of the cable as shown in figures. 2 and 3.

The value of magnetic field at the tape surface determines the value of critical current. First, it is necessary to calculate the value of magnetic field acting on the tape surface when the transport current flow to the both layers in case of same and opposite winding directions.

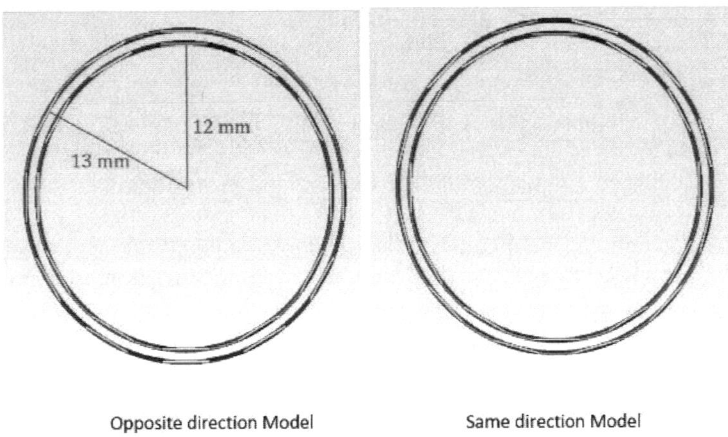

Figure 1. The cross section of two-layer cable with widing on the opposite and same directions.

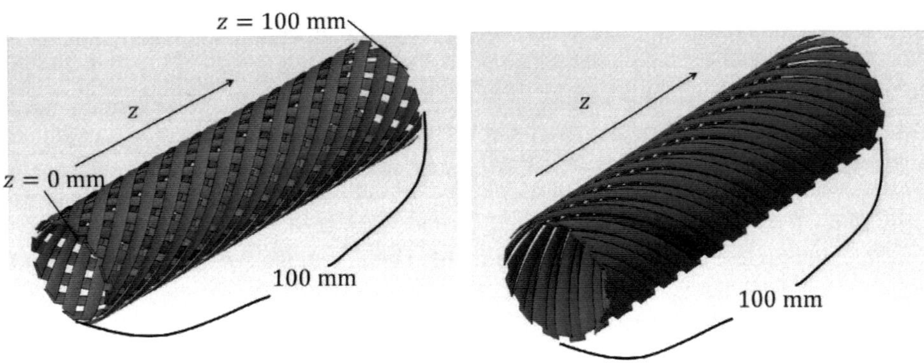

Figure 2. The calculation model with opposite direction winding (left) and same direction winding (right).

Figure 4 shows the calculated result of position dependence of perpendicular (*c*-axis) component of magnetic field to the tape surface for the cases when superconducting tape layers winded in the same and opposite directions. It is found that the value of perpendicular component of magnetic field is less in the case of superconducting tape layers winded in the same direction. It means that the value of critical current is higher for the case of same winding direction.

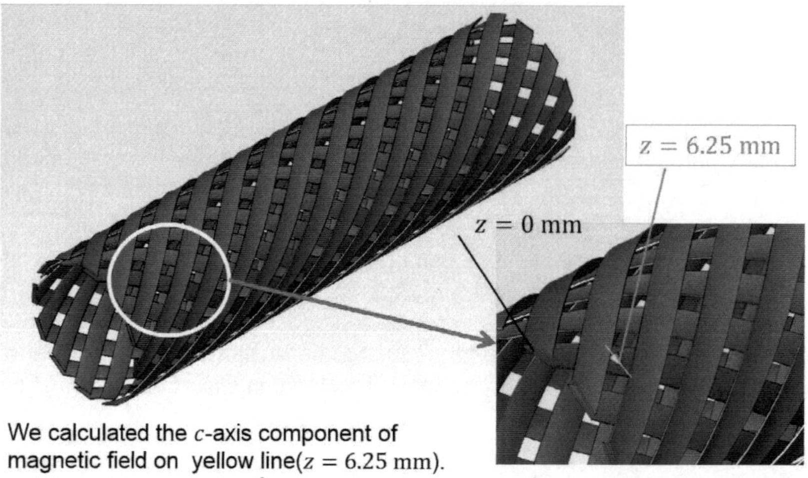

Figure 3. Position of $z = 0$ mm(0^0) and $z = 6.25$ mm(7.5^0) in calculation model of opposite winding direction.

For the case when the winding performs in opposite direction, the value of magnetic field on the surface of HTS tape changes as a function of z-axis, and it takes minimum value as same as the case of the same direction winding at $z = 6.25$ mm. The value of magnetic field takes maximum value and is two times larger than that in the minimum case.

Figure 5 shows the position dependence of perpendicular (c-axis) component of magnetic field for various position of cable cross section for the case of opposite direction winding. The curve 0^0 corresponds the cross section when the tapes of different layer touches each other as shown in figure 3 at $z = 0$ mm. The curve 7.5^0 corresponds position $z = 6.25$ mm in figure 3. 360^0 is corresponds to position of $z = 100$ mm. It is found that the position dependence of magnetic field changes periodically with the angle of 7.5^0

It is necessary to calculate the value of critical current in two-layer cable in case of winding in the same and opposite directions. For the case of winding in the same direction the distribution of magnetic field along the z-axis is homogeneous. The value of critical current is possible to determine through the average voltage on the cable according criterion of electrical field in the cable, $E_0 = 10^{-4}$ V/m.

The perpendicular magnetic field dependence of critical current density of the YBCO tape from the data of reference [5] and approximation function are shown in figure 6.

The critical current of straight tape in two-layer cable with winding in the same direction is possible to calculate from the figure 4 (lower curve, corresponding same direction winding) and figure 6. The distribution of normalized critical current in the width $J(B(x))$ is shown in figure 7 (upper curve). The important parameter is the average value of critical current density, and is 0.938. It means that the value of critical current is 0.938 from the maximum value in case of absence of magnetic field acting on the tapes.

The measurement of critical current of single tape and the cable with same length can give the different results because the length of the tape of the cable is longer compare with the straight tape. In our case, the cable is inclined with angle $\varphi = 37°$ (because twist pitch = 100 mm, circle length = $2 \cdot \pi \cdot R = 2 \cdot \pi \cdot 12$ mm = 75 mm). The length of tape for 1 m of the cable will be $\frac{1}{\tan(\varphi)} = 1.327$ m.

IOP Conf. Series: Journal of Physics: Conf. Series **897** (2017) 012017 doi:10.1088/1742-6596/897/1/012017

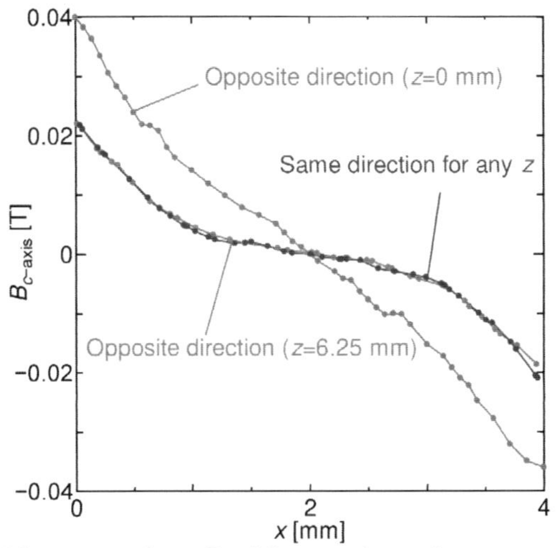

Figure 4. Position dependence of perpendicular (c-axis) component of magnetic field to the tape surface for the cases when superconducting tape layers winded in the same and opposite directions.

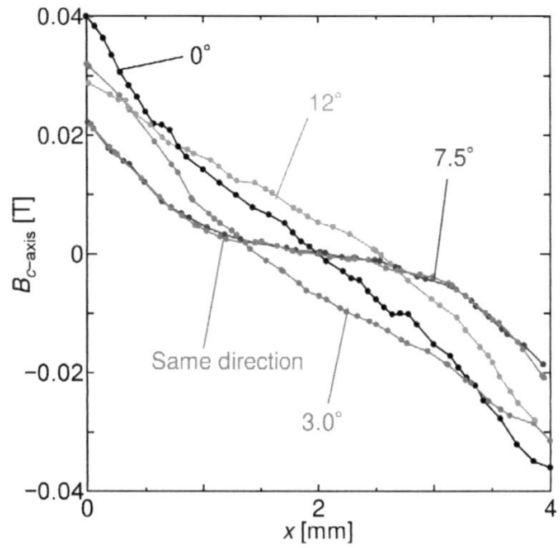

Figure 5. The position dependence of perpendicular component of magnetic field for different cable cross section for the case of opposite direction winding.

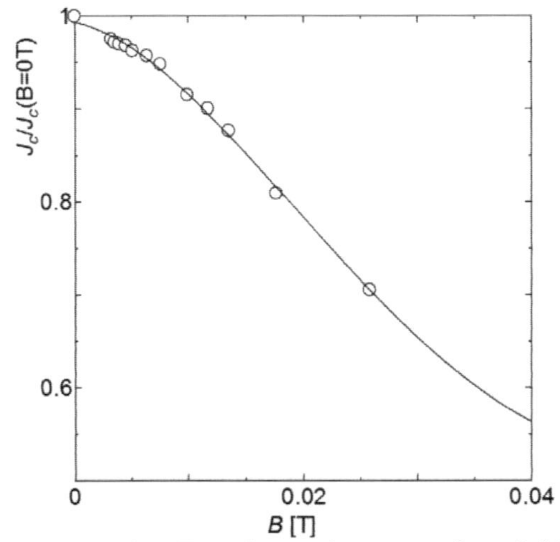

Figure 6. Experimental magnetic field dependence of critical current density of the YBCO in the perpendicular magnetic field and approximation function [5].

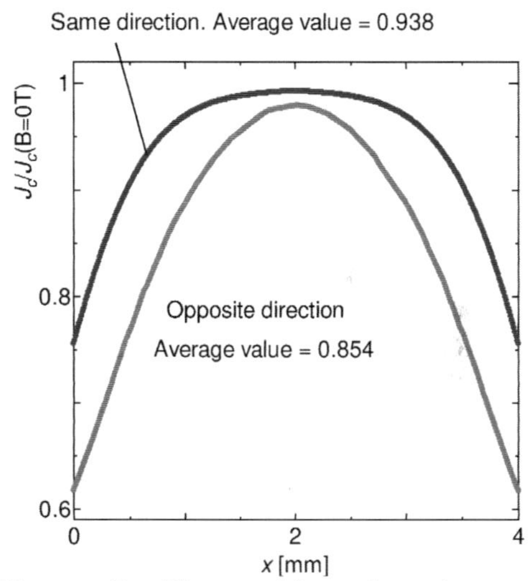

Figure 7. The position dependence of normalized critical current density.

The voltage on the cable for 1 m is $V = E_0 \left(\frac{I}{I_c}\right)^n$, where I is transport current, I_c is critical current which is determined when the electric field of the cable becomes $E_0 = 10^{-4}$ V/m, the coefficient n-value is $n = 10$ to 20, respectively. The length of the cable increases $\frac{1}{\tan(\varphi)} = 1.327$ times. This is equal to the ratio of length of tape in the cable $l_{\text{tape in cable}}$ and that of single tape $l_{\text{tape alone}}$, $\frac{l_{\text{tape in cable}}}{l_{\text{tape alone}}}$. Then the voltage taking into account the changing of length is

$$V = \frac{l_{\text{tape in cable}}}{l_{\text{tape alone}}} V_0 \left(\frac{I}{I_c}\right)^n, \tag{1}$$

where V_0 is the voltage for the case that tape length is cable length without twisting. The length of tapes is larger for the case of twisting, and the voltage is larger. To obtain the same voltage $V = V_0$ it is necessary to decrease current. In this case, the value of critical current of the cable in case of longer tape is:

$$I_{\text{c cable}} = I_{c0} \cdot \sqrt[n]{\frac{l_{\text{tape in alone}}}{l_{\text{tape cable}}}} = I_{c0} \sqrt[n]{\cos\varphi} = I_{c0} \cdot \sqrt[10\div 20]{0.8} = 0.98 \div 0.99\, I_{c0}. \tag{2}$$

Therefore, there is no difference between single tape and cable from point of view critical current measurement in longer tape.

For the case of cable winded in the opposite directions, the situation is more complicated, since the magnetic field along the tape is not homogeneous and there exists the underloaded and overloaded areas. Therefore, for simplicity, the current is assumed to flow parallel each other and total voltage of the tape can be obtained from the integral of voltage drop. Then, critical current is determined at the electric field reaches to the electric field criterion, $E_0 = 10^{-4}$ V/m. The total voltage is given by

$$V(x) = \int_{z=0}^{z=1000mm} \left(\frac{I(x)}{I_c(x,z)}\right)^n U_0 \cdot dz = V_0. \tag{3}$$

The value of local critical current depends on the coordinates from the edge of the tape (x) and coordinate of cross section (z). For each value of x the voltage $V(x)$ was calculated and the transport current of each x was taken to obtain the value $V(x) = V_0$.

The dependence of transport current as a function of x is shown in figure 7 (lower curve). This is the position distribution of critical current in the tape. The average value of critical current is 0.854. It is about 10% less than for the case of same direction winding. Although the effects of longitudinal magnetic field are not taken into account, this effect make the profit of same direction winding.

It is possible to conclude that the same direction winding is more profitable from the view point of the critical current, since the perpendicular component of magnetic field is smaller in the case of same direction winding.

3. Conclusion

In the present study, the comparison of the critical current of same and opposite direction winding of two-layer cable with same twist pitch was performed. The calculation of the magnetic field distribution was made with finite element method. First, the distribution of magnetic field on the HTS tapes of the cable was calculated. And then, the critical current was calculated in both cases. It is found that the same direction winding is at least 10% better than opposite direction winding for critical current of the cable. The future research will concern the experimental research with this type of cable.

References

[1] Vyatkin V S, Tanabe K, Wada J, Kiuchi M, Otabe E S and Matsushita T 2013 Calculation of critical current in DC HTS cable using longitudinal magnetic field effect *Physica C* **494**, 135

[2] Vyatkin V S, Kiuchi M, Otabe E S, Ohya M and Matsushita T 2015 Current- carrying capacity of single layer cable using superconducting Bi-2223 tapes in parallel magnetic field *Supercond. Sci. Technol.* **28** 015011 (3pp)

[3] Yamaguchi S, Hamabe M, Yamamoto I, Famakinwa T, Sasaki A, Iiyoshi A and Kawamura K 2008 Research activities of DC superconducting power transmission line in Chubu University. *Journal of Physics: Conference Series* **97** 012290

[4] JMAG https://www.jmag-international.com/

[5] Matsui V I, Flis V S, Moskaliuk V O, Kasatkin A L, Skoryk N A and Svechnikov V L 2015 Current-Carrying Abilities of Nano-Structured HTS Thin Films *Journal of Nanoscience and Nanoengineering* **1** 38

IOP Conf. Series: Journal of Physics: Conf. Series **897** (2017) 012018 doi:10.1088/1742-6596/897/1/012018

Study on Separation of Structural Isomer with Magneto-Archimedes method

T Kobayashi[1], T Mori[1], Y Akiyama[1], F Mishima[2] and S Nishijima[1]

[1]Graduate school of engineering, Osaka University, 2-1 Yamadaoka, Suita, Osaka, 565-0871, Japan

[2]Department of applied nuclier technology, Fukui University of technology, 3-6-1 Gakuen, Fukui-City, Fukui, 910-8505, Japan

E-mail: t-kobayashi@qb.see.eng.osaka-u.ac.jp

Abstract. Organic compounds are refined by separating their structural isomers, however each separation method has some problems. For example, distillation consumes large energy. In order to solve these problems, new separation method is needed. Considering organic compounds are diamagnetic, we focused on magneto-Archimedes method. With this method, particle mixture dispersed in a paramagnetic medium can be separated in a magnetic field due to the difference of the density and magnetic susceptibility of the particles. In this study, we succeeded in separating isomers of phthalic acid as an example of structural isomer using $MnCl_2$ solution as the paramagnetic medium. In order to use magneto-Archimedes method for separating materials for food or medicine, we proposed harmless medium using oxygen and fluorocarbon instead of $MnCl_2$ aqueous solution. As a result, the possibility of separating every structural isomer was shown.

1. Introduction

Organic compounds are used in many fields such as industrial products, foods and medicines. They are refined by separating their structural isomers contained in the raw materials. The separation methods are distillation, crystallization and extraction. However, distillation consumes the large amount of energy. The crystallization and the extraction have a problem in the low separation accuracy [1]. These problems could be remarkable in separating structural isomers, etc. that have very similar physical properties. Therefore, new separation method is needed from the viewpoint of environmental conservation.

In this study, we focused on the difference of magnetic susceptibility and density between organic compounds and their structural isomers, that is, we used a magnetic separation method. Considering organic compounds are diamagnetic, we proposed magneto-Archimedes method which is effective for separation of diamagnetic mixture. There are some reports studying magneto-Archimedes separation for metals, foods or plastics [2, 3, 4]. However, the separation of the organic compound and its structural isomers has not been reported before. In this work, we chose isomers of phthalic acid (figure 1) as the separation objects. Each of them is white particle whose size is about 50 μm.

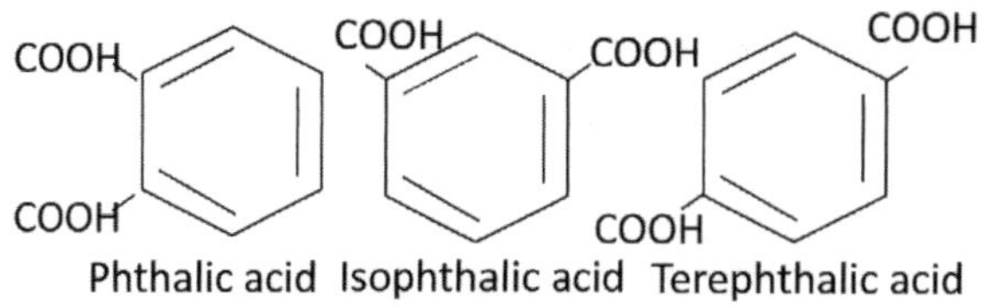

Figure 1. Structural isomers of phthalic acid.

2. Magneto-Archimedes method

2.1. Magneto-Archimedes effect [5]

Particle mixture dispersing in the medium can be separated in the magnetic field with magneto-Archimedes method, using the difference of levitation or sedimentation height of each particle. Levitation or sedimentation is induced by magneto-Archimedes effect.

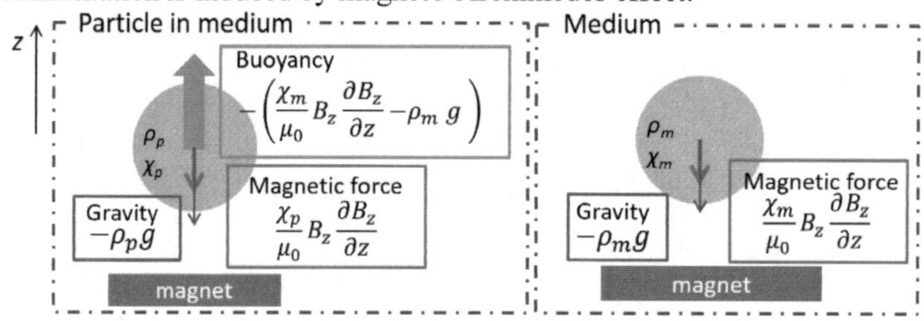

Figure 2. Diagram of magneto-Archimedes effect.

In figure 2, z axis was set to vertical direction where upward is positive. The resultant force acting on the particle in the medium which is placed in the magnetic field, F_z, is the sum of the magnetic force, gravity force and buoyancy. Based on the Archimedes' principle, buoyancy is the reaction force of the resultant force acting on the medium which has the same volume as the particle. Thus, F_z can be expressed as equation (1). Here, χ (-), μ_0 (N/A^2), B (T), ρ (kg/m^3) and g (m/s^2) are magnetic susceptibility, space permeability, magnetic flux density, density and gravitational acceleration, respectively. The subscripts p and m correspond to particle and medium. Magnetic force acting on the particle is expressed in the first term. Considering $\partial B_z/\partial z$ is negative, magnetic force acting on the paramagnetic particle (χ_p> 0) is negative, which corresponds to attractive force. On the other hand, the force acting on the diamagnetic particle (χ_p< 0) is positive, which corresponds to repulsive force.

$$F_z = \frac{\chi_p}{\mu_0} B_z \frac{\partial B_z}{\partial z} + \left(-\rho_p g\right) + \left\{-\left(\frac{\chi_m}{\mu_0} B_z \frac{\partial B_z}{\partial z} - \rho_m g\right)\right\} \tag{1}$$

Since paramagnetic medium is usually used in magneto-Archimedes method, χ_m is positive. Since $\partial B_z/\partial z$ is negative, and then the magnetic force acting on the medium is negative, corresponding to the third term in the parentheses. Therefore, buoyancy acting on the particle placed in the magnetic field is positive and becomes larger than that without the magnetic field, which results in the particle levitation. This effect is called magneto-Archimedes effect.

2.2. The mechanism of magneto-Archimedes method

Here, the particle levitation mechanism of magneto-Archimedes method is explained. When buoyancy is larger than the sum of magnetic and gravity force ($F_z \geq 0$), the particle levitates. When the three forces balance ($F_z = 0$), the particle keeps levitation at certain height. Because $\chi_p < \chi_m$ in this work, the levitation condition ($F_z \geq 0$) can be shown in equation (2).

$$B_z \frac{\partial B_z}{\partial z} \leq \frac{\rho_p - \rho_m}{\chi_p - \chi_m} \mu_0 g \tag{2}$$

Usually, χ_p of the diamagnetic material is negative and its magnitude is much smaller than χ_m, therefore χ_p -χ_m of equation (2) can be approximated as -χ_m. When equation (2) is satisfied, the relationship between ρ_p and $B_z\partial B_z/\partial z$ can be shown in figure 3(a).

$B_z\partial B_z/\partial z$ required for levitation of each particle, denoted as 1 and 2, having density of ρ_1 and ρ_2 are calculated and presented in the figure 3(b). Figure 3(b) shows $B_z\partial B_z/\partial z$ as a function of height from the magnet used in this experiment. According to figure 3(b), particle 1 and 2 levitate at h_1 and h_2 respectively. Using the difference of levitation heights, each particle can be separated.

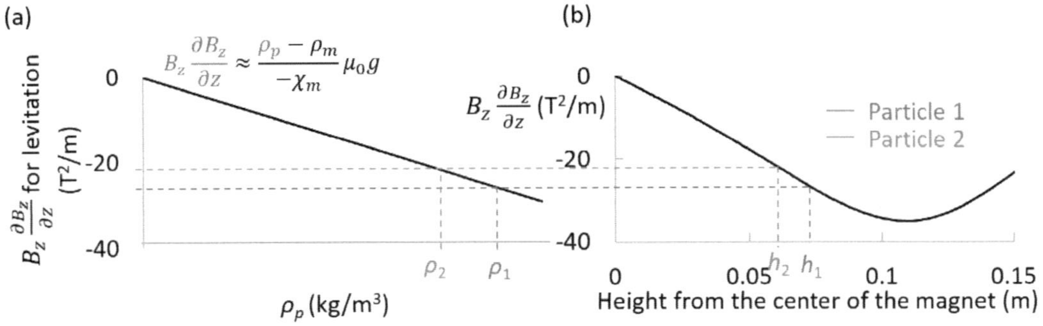

Figure 3. Mechanism of separation. Relation between density of particle and $B_z\partial B_z/\partial z$ for levitation is shown in (a). $B_z\partial B_z/\partial z$ distribution against height from magnet center is shown in (b).

When the concentration or composition of the medium is changed, the right-hand side of equation (2) is also changed. Thus, the inclination and the intercept of figure 3(a) depends on the medium, which affects the levitation height of each particle. In addition, the profile of figure 3(b) depends on the magnet, which also affects the levitation height. Effective separation is possible when the appropriate medium and magnet are chosen.

2.3. Flowchart for Magneto-Archimedes method

Based on above discussion, we proposed the procedure of Magneto-Archimedes separation was shown in figure 4. It consists of determining the physical property of separation objects, choice of medium, calculation of $B_z\partial B_z/\partial z$ for levitation, choice of magnet, estimation of levitation height, and then separation experiment. The separation conditions in this study were determined based on the flowchart presented in figure 4.

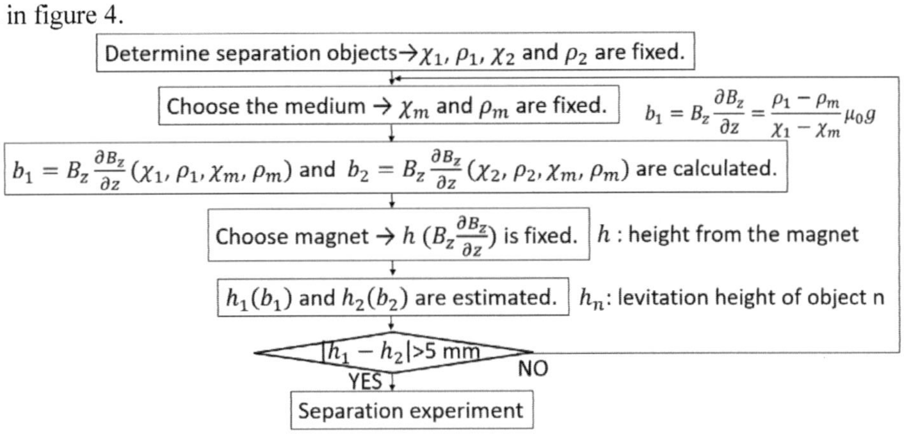

Figure 4. Flowchart for magneto-Archimedes method.

3. Separation of isomers of phthalic acid

3.1. Choice of the medium and magnet

The density and susceptibility of isomers of phthalic acid are shown in table 1[6, 7, 8].

Table 1. Density and susceptibility of the structural isomers.

	Density (kg/m^3)	Susceptibility (-)
Phthalic acid	1590	-1.00×10^{-5}
Isophthalic acid	1540	-9.85×10^{-6}
Terephthalic acid	1510	-9.53×10^{-6}

We chose $MnCl_2$ aqueous solution which shows paramagnetism as the medium. Due to high solubility in water, the density and susceptibility of $MnCl_2$ solution can be controlled by its concentration. Figure 5 shows the relation between concentration of $MnCl_2$ solution and $B_z\partial B_z/\partial z$ for levitation of each particle which are calculated by equation (2) and table 1.

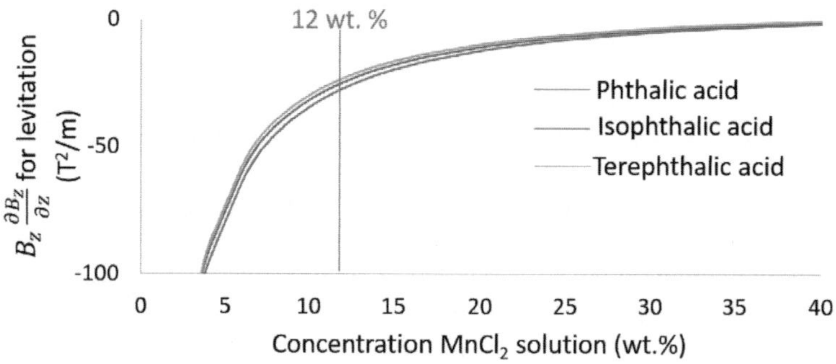

Figure 5. Relation between concentration of $MnCl_2$ solution and $B_z\partial B_z/\partial z$ for levitation of each particle.

Figure 5 shows that the difference of $B_z\partial B_z/\partial z$ for levitation between three structural isomers, demonstrating the largest difference at 10-15 wt. % $MnCl_2$ solution. Thus, 12 wt. % $MnCl_2$ solution, whose density was 1100 kg/m^3 and magnetic susceptibility was 2.10×10^{-4}, was chosen as the medium. However, even in case 12 wt. % $MnCl_2$ solution was used, the difference of $B_z\partial B_z/\partial z$ for levitation among three acids was too small to separate with permanent magnet. Therefore, superconducting solenoidal magnet shown in figure 6 (model number: JMTD-10T100E3, maximum magnetic flux density: 10 T) was used.

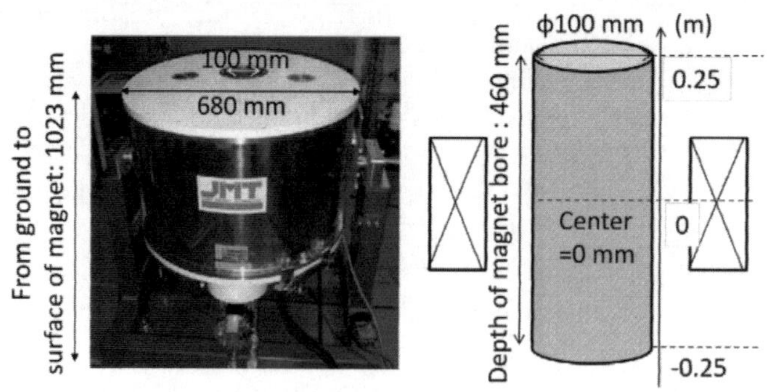

Figure 6. Superconducting solenoidal magnet.

3.2. Estimation of levitation heights

Figure 7(a) shows $B_z\partial B_z/\partial z$ for levitation of the particle as a function of ρ_p with 12 wt. % $MnCl_2$ solution. In this case, each isomer gets on the same line because (χ_p -χ_m) of each isomer can be approximated as "-χ_m", in other words, χ_m is much larger than χ_p as shown in table 1. Figure 7(b) shows $B_z\partial B_z/\partial z$ distribution against height from the center of magnet which generates 3 T in the maximum magnetic flux density. Based on figure 7(a) and table 1, $B_z\partial B_z/\partial z$ for levitation of each particle was estimated by figure 7(a) and levitation height was estimated by figure 7(b). Each particle was estimated to levitate at the height in the order of terephthalic (0.059 m), isophthalic (0.064 m) and phthalic acid (0.071 m).

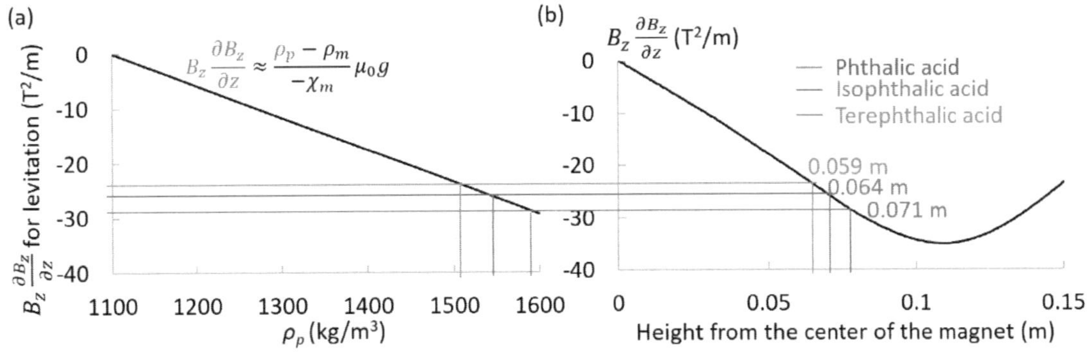

Figure 7. Estimation of levitation height of each object. $B_z\partial B_z/\partial z$ for levitation as a function of density of particle is shown in (a). $B_z\partial B_z/\partial z$ distribution against height from magnet center is shown in (b).

3.3. Separation experiment

Based on above estimation, we conducted separation experiment. Firstly, 12 wt. % $MnCl_2$ solution and mixture of separation objects were put into the glass vessel and installed near the center of superconducting solenoidal magnet. As a result, due to the density difference between the medium and each particle, all the isomers settled to the bottom. Then, the vessel was elevated slowly and each isomer levitated at the different height in the order of terephthalic (0.074 m), isophthalic (0.085 m) and phthalic acid (0.092 m). Figure 8 shows the result. This phenomenon was simulated by the calculation with satisfactory accuracy, indicating that we succeeded in separating isomers of phthalic acid.

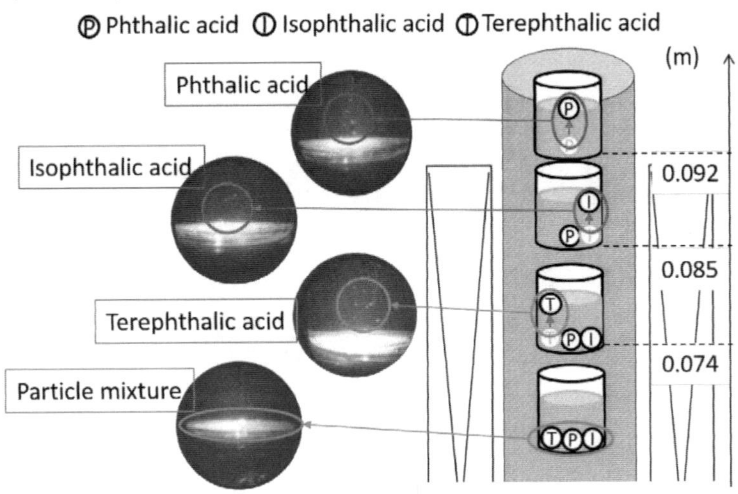

Figure 8. Separation experimental result with 12 wt. % $MnCl_2$ solution.

4. Separation without heavy metal solution

4.1. Choice of the medium and magnet

Foods and medicines are also the subjects of the interest. Paramagnetic medium including heavy metals such as $MnCl_2$ solution should not be suitable for separation of foods or medicines, because heavy metal is harmful to human body. In order to separate materials of foods or medicines, paramagnetic medium without heavy metals should be examined. Oxygen gas is paramagnetic, however the density is very small and magneto-Archimedes effect is hard to be observed.

We focused on oxygen dissolved fluorocarbon as new medium. Fluorocarbon has carbon-fluorine bonds. Due to the bond, intermolecular force is smaller than that of hydrocarbon, and oxygen gas can be dissolved into fluorocarbon easily. Therefore, when high pressure oxygen gas is dissolved in fluorocarbon, it becomes paramagnetic [9]. Moreover, many fluorocarbons is volatile, so oxygen dissolved fluorocarbon is suitable for foods and medicines. Physical property of fluorocarbon used in this study ($C_4F_9OCH_3$) is shown in table 2 [10].

Table 2. Physical property of fluorocarbon in this study.

Density (kg/m^3)	Susceptibility (-)	Boiling point (℃)	Melting point (℃)	Oxygen solubility (mL/L)
1520	-7.75×10^{-6}	61	-135	300

Magnetic susceptibility of oxygen dissolved fluorocarbon was calculated, based on the assumption that oxygen gas was ideal gas which follows Henry's law. The calculation result is shown in figure 9. Magnetic susceptibility increased linearly against the pressure of oxygen gas and became positive at 1.5 MPa.

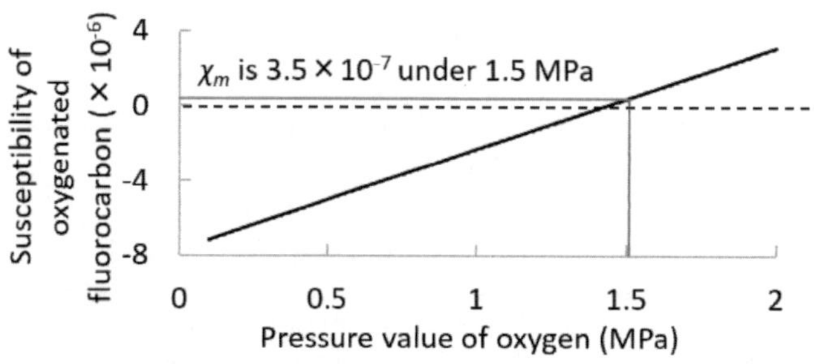

Figure 9. Relation between pressure of oxygen gas and the calculated susceptibility of oxygen dissolved fluorocarbon.

In this chapter, pellets (φ3×1.5 mm) of isomers of phthalic acid prepared with tablet forming device were separated. The density of each pellet measured with an analytical valance (model number: AUX-220) with a specific gravity measurement kit (model number: SMK-401) is shown in table 3. According to table 1 and 3, the magnitude relation of the density of each isomer in pellet form differs from that in powder form. That may be caused by air included in each pellet, but the clear reason is currently under investigation.

Table 3. Density and susceptibility of each pellet.

	Density (kg/m^3)
Phthalic acid	1560
Isophthalic acid	1500
Terephthalic acid	1570

Based on table 3, $B_z\partial B_z/\partial z$ for levitation as a function of density of particle at 1.5 MPa oxygen dissolved fluorocarbon is calculated and presented in figure 10(a). In this case, isophthalic acid was

IOP Conf. Series: Journal of Physics: Conf. Series **897** (2017) 012018 doi:10.1088/1742-6596/897/1/012018

excluded because it was estimated to float in the medium without magnetic field due to the difference of density between isophthalic acid and fluorocarbon. Moreover, each isomer does not get on the same line. The difference between figure 7(a) and figure 10(a) is brought about by the difference of (χ_p -χ_m) in equation (2). In this case, (χ_p -χ_m) cannot be approximated as "-χ_m". This is because magnitude of χ_m is smaller than that of χ_p as shown in figure 9 and table 3. From figure 10(a), $B_z\partial B_z/\partial z$ for levitation of each isomer was calculated as -48.1 and -60.0 T^2/m respectively for phthalic and terephthalic acid. The difference between -48.1 and -60.0 T^2/m was very small to separate pellets with permanent magnet. On the basis of the calculation, superconducting solenoidal magnet shown in figure 6 was also used.

4.2. Estimation of levitation heights

Figure 10(b) shows the distribution of $B_z\partial B_z/\partial z$ against the height from the center of the magnet which generates 8 T at the center of the magnet. From figure 10, each isomer was estimated to levitate at 0.192 m for terephthalic and 0.199 m for phthalic acid.

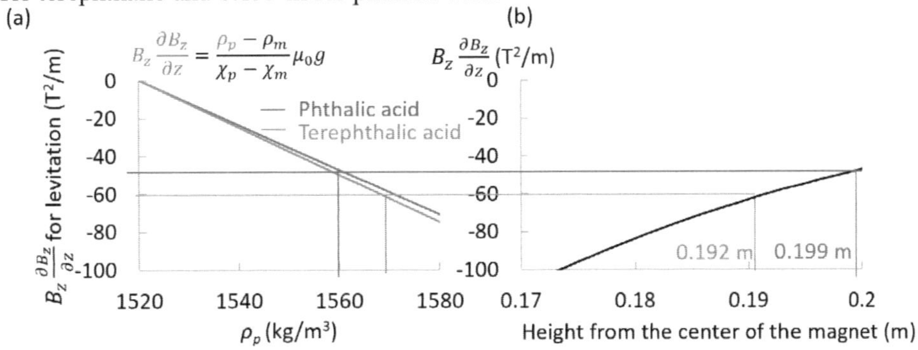

Figure 10. Estimation of levitation height of each pellet. $B_z\partial B_z/\partial z$ for levitation as a function of density of particle is shown in (a). $B_z\partial B_z/\partial z$ distribution against height from magnet center is shown in (b).

4.3. Separation experiment

Firstly, the pressure resistant vessel consisting of glass and polycarbonate vessels shown in figure 11(a) and (b) are prepared, and fluorocarbon and three different pellets were encapsulated in the vessel. Then, isophthalic acid floated on the surface of medium, whereas other two pellets settled at the bottom of vessel. After that, 1.5 MPa oxygen gas was fed into the vessel as shown in figure 11(c). Then the vessel was put at 0.12 m height from the center of the magnet. Each pellet was expected to levitate at different height as shown in figure 11(d). The experimental photograph is shown in figure 11(e). Each pellet levitated at different height from the center of magnet in the order of terephthalic (0.19 m), phthalic (0.20 m) and isophthalic acid (surface of medium) similar to the estimation as figure 11(d). The result indicates the possibility of separation of isomers of phthalic acid with oxygen dissolved fluorocarbon.

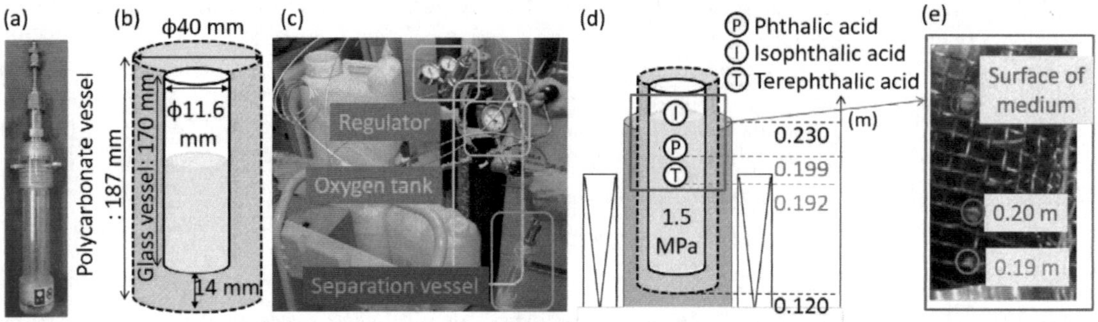

Figure 11. Result of separation experiment. Appearance and overview of separation vessel are shown in (a) and (b) respectively. Appearance of pressurization to fluorocarbon in the vessel is shown in (c). Calculation and experimental result of separation of pellets are shown in (d) and (e) respectively.

5. Conclusion

In this study, we developed new method to separate structural isomers, and tried to separate isomers of phthalic acid by magneto-Archimedes method. Based on calculation for effective separation, 12 wt. % $MnCl_2$ solution as a medium and superconducting solenoidal magnet generating 3 T were used. As a result, isomers of phthalic acid were separated as predicted by calculation.

Moreover, considering the application of magneto-Archimedes method for foods and medicines, we proposed oxygen dissolved fluorocarbon as the medium without heavy metals. 1.5 MPa oxygen dissolved fluorocarbon and superconducting solenoidal magnet generating 8 T were used, based on the designing for the effective separation. As a result, we succeeded in separating pellets of isomers of phthalic acid. In progress, we are trying to separate powdered isomers of phthalic acids with oxygen dissolved fluorocarbon.

tsman Corporation, *the Safety Data Sheet of the AROCY® 35000 CH*

References

[1] National institute of advanced industrial science and technology (AIST) 2006 *AIST TODAY* **6** 16
[2] Matsuura Y and Miura O 2016 *IEEE Transactions on Applied Superconductivity* **26** 4
[3] Maki S and Hirota N 2014 *Journal of Food Engineering* **120** 31-36
[4] Ueda Y, Mishima F, Akiyama Y and Nishijima S 2014 *IEEE Transactions on Applied Superconductivity* **24** 3
[5] Liu S, Leaper M and Miles J N 2014 *Powder Technology* **261** 71-77
[6] Wako pure chemical industries, Ltd. 2014 *safety data seat of phthalic acid*
[7] Wako pure chemical industries, Ltd. 2016 *safety data seat of isophthalic acid*
[8] Wako pure chemical industries, Ltd. 2016 *safety data seat of terephthalic acid*
[9] Mishima F, Akiyama Y and Nishijima S 2014 *IEEE Transactions on Applied Superconductivity* **24** 3
[10] Ohmori Y and Kajiyama T 1985 *The chemical society of Japan* **10** 1897-1903

Complete Fabrication of a Traversable 3 μm Thick NbN Film Superconducting Coil with Cu plated layer of 42m in Length in a Spiral Three-Storied Trench Engraved in a Si Wafer of 76.2 mm in Diameter Formed by MEMS Technology for a Compact SMES with High Energy Storage Volume Density

Yasuhiro Suzuki[1], Nobuhiro Iguchi[2], Kazuhiro Adachi[1], Akihisa Ichiki[1], Tatsumi Hioki[1], Che-Wei Hsu[3], Ryoto Sato[3], Shinya Kumagai[3], Minoru Sasaki[3], Joo-Hyong Noh[4], Yuuske Sakurahara[4], Kyohei Okabe[4], Osamu Takai[4], Hideo Honma[4], Hideo Watanabe[5], Hitoshi Sakoda[5], Hiroaki Sasagawa[5], Hideyuki Doy[5], Shuliang Zhou[6], H. Hori[6], Shigeaki Nishikawa[6], Toshihiro Nozaki[6], Noriaki Sugimoto[3,7], Tomoyoshi Motohiro[1,2,3,7]

1. Materials Science and Energy Engineering Division, Green Mobility Research Institute, Institutes of Innovation for Future Society, Nagoya University, Furo-cho, Chikusa-ku, Nagoya,464-8603, Japan
2. Graduate School of Engineering, Nagoya University, Furo-cho, Chikusa-ku, Nagoya,464-8603, Japan
3. Graduate School of Engineering, Toyota Technological Institute, Hisakata 2-12-1, Tempaku-ku, Nagoya 468-8511, Japan
4. Materials & Surface Engineering Research Institute, Kanto-Gakuin University, 1-1-1 Fukuura, Kanazawa-ku, Yokohama 236-0004, Japan
5. D-process Inc., 2-8-22 Yamato-minami, Yamato-shi, Kanazawa, 242-0006, Japan
6. AISIN SEIKI Co. Ltd., Advance Square Karia, 1-1-1, Aioi-cho, Kariya, Aichi, 448-8605, Japan
7. TOYOTA Central R&D Labs., Inc., Nagakute, Aichi, 480-1192, Japan

motohiro@gvm.nagoya-u.ac.jp

Abstract. Based on the concept of a novel approach to make a compact SMES unit composed of a stack of Si wafers using MEMS process proposed previously, a complete fabrication of a traversable 3μm thick NbN film superconducting coil lined with Cu plated layer of 42m in length in a spiral three-storied trench engraved in and extended over a whole Si-wafer of 76.2 mm in diameter was attained for the first time. With decrease in temperature, the DC resistivity showed a metallic decrease indicating the current pass was in the Cu plated layer and then made a sudden fall to residual contact resistance indicating the shift of current pass from the Cu plated layer to the NbN film at the critical temperature T_c of 15.5K by superconducting transition. The temperature dependence of *I-V* curve showed the increase in the critical current with decrease in the temperature and the highest critical current measured was 220 mA at 4K which is five times as large as that obtained in the test fabrication as the experimental proof of concept presented in the previous report. This completion of a one wafer superconducting NbN coil is an indispensable step for the next proof of concept of fabrication of series-connected two wafer coils via superconductive joint which will read to series connected 600 wafer coils finally, and for replacement of NbN by high T_c superconductor such as $YBa_2Cu_3O_{7-x}$ for operation under the cold energy of liquid hydrogen or liquid nitrogen.

1. Introduction

Compensation for fluctuating output power of a renewable energy source using an energy storage system is indispensable for extending the usage of renewable energy. N. Ishida et al. proposed a hybrid energy storage system composed of a hydrogen storage system and an electricity storage system [1]. More concretely, the proposed system has a high pressure hydrogen gas bus line and a electricity bus line, energy conversion between two bus lines are done using a fuel cell and a water electrolysis cell. The hydrogen bus line is connected to dispensers for fueling fuel cell vehicles in the downstream with a high pressure hydrogen gas tank as a reservoir, and connected to liquid hydrogen tank via a heat exchanger for conversion between liquid and gas bi-directionally. Refills of liquid hydrogen are obtained basically as a by-product from industrial plants and complexes and transported by tank trucks .The electricity bus line is connected to renewable energy sources such as mega solar plants and wind farms in the upstream and also connected to local grid for civil and industry use in the downstream. Since the fluctuations of the output powers from mega solar plants and wind farms are very irregular involving fairly amount of high frequency components, N. Ishida et al. recommended usage of SMES utilizing low temperature of liquid hydrogen rather than secondary battery systems which are not good at rapid charge and discharge operation[1]. Since secondary batteries operate ionic transportation and impregnation into electrodes as chemical reactions at charge and discharge, rapid and irregular charge and discharge operation may cause degradation of the electrodes and problems in the long term operations. On the contrary, SMES is good at rapid charge and discharge operations and quite stable for long term use based on purely electronic transportation. Figure 1 displays a possible future territorially distributed H_2 and electricity fueling station based on a modified redraw of the concept of N. Ishida et al. on compensation for fluctuating output power of a renewable energy source using a hybrid energy storage system [1]. In this figure, electric power supply to electric vehicles and hydrogen supply as solid metal hydrides is also added for future possibility.

However, N. Ishida et al. could not construct the experimental proof of their concept design using a SMES system because it is not easy to get a SMES system and used capacitors as a substitute since capacitors are easily commercially available and also good at rapid charge and discharge operation if the required capacity is not so large in the experimental proof of concept [1]. A capacitor is an electrical power storage system in which electrical power is stored as energy of electrostatic field, and SMES is an electrical power storage system in which electrical power is stored as energy of magnetic field. In this sense, SMES is a counterpart of capacitor in electromagnetics. However, conventional SMESs are gigantic systems constructed as a made-to-order product by heavy electrical industry, while capacitors

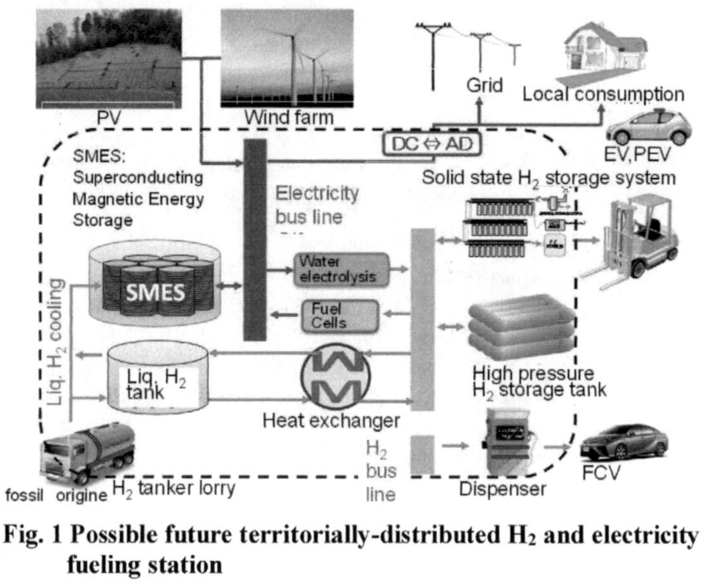

Fig. 1 Possible future territorially-distributed H_2 and electricity fueling station

are mass-produced devices and scale-upped by combining units of mass-produced devices. The concept of a novel approach to make a compact SMES unit composed of a stack of Si wafers using MEMS process proposed in our previous report may bring the SMES to be a real counterpart of capacitor to be able to be used in the proposal of N. Ishida et al. mentioned above both in the small scale experimental proof and practically realized system for storage of renewable energy in a large scale [2]. In this paper, the first complete fabrication of a traversable 3μm thick NbN film superconducting coil lined with Cu plated layer of 42m in length in a spiral three-storied trench engraved in and extended over a whole Si-wafer of 76.2 mm in diameter is reported based on our previous proposal[2].

2. Experimental

2.1. New design and fabrication of a spiral trench on a Si wafer for reactive sputter deposition of thicker NbN superconducting film

Figure 2 (a) illustrates the cross-sectional dimension of trenches in Fig.4 in the previous report [2]. To increase energy storage capacity, enlargement of the cross sectional area is requested. In addition to this, v-shaped cross-section as shown in Fig. 2(b) is advantageous for thicker NbN film formation by incoming sputtered particles from above. Widening of trench width decreases number of turns of the spiral coil which is proportional to inductance of the coil and hence the energy storage capacity. However, the critical superconducting current increases in proportion to the square of the trench width, and the energy storage capacity increases in proportion to the square of the critical current. Therefore, widening of trench width contributes increase in the energy storage capacity. Fig.2(c) is actually attained cross-section of the trench by the MEMS process as shown in Fig.3 including repeated reactive ion etching (RIE) using SF_6 plasma and C_4F_8 plasma alternatively (Bosh process) as was done in the previous study[2]. The overall shape is v-shaped but it is composed of two sets of two terraces, one bottom and two sets of three perpendicular walls forming four storied fluvial terrace-like structure. The cross-sectional area of the trench shown in Fig.2 (c) is 17 times as large as that in Fig.2 (a). Figure 4 shows a Si wafer of 76.2 mm in diameter after fabrication of

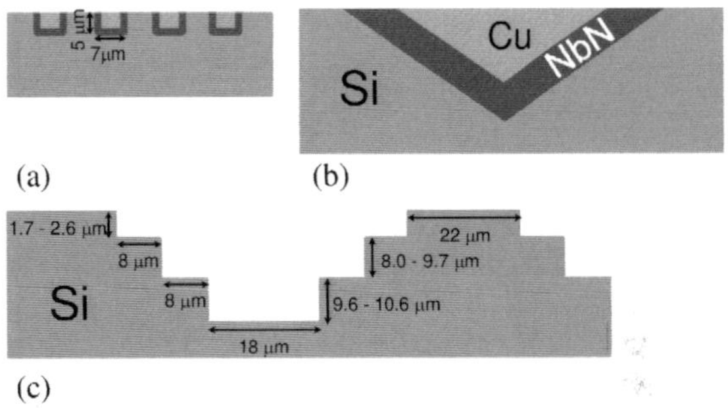

Fig.2 Cross-sectional designs and dimensions of trenches in (a) previous study, (b) ideal improvement, (c) actually realized one in this study

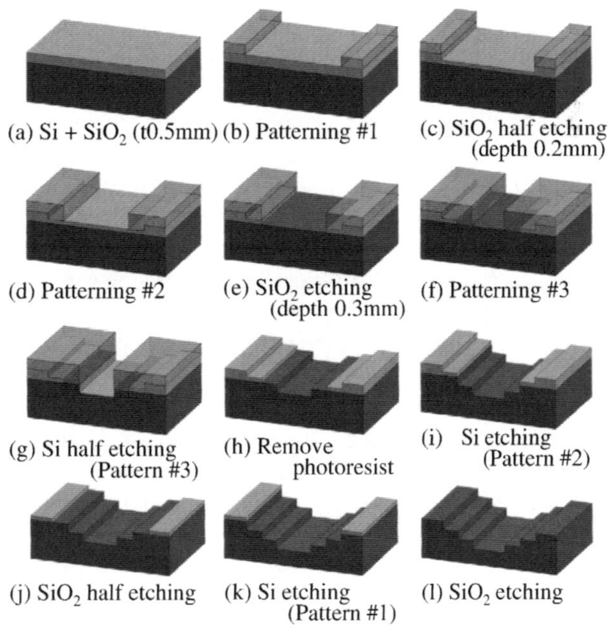

Fig.3 Si MEMS process to form a trench of four storied fluvial terrace-like structure.

IOP Conf. Series: Journal of Physics: Conf. Series **897** (2017) 012019 doi:10.1088/1742-6596/897/1/012019

newly designed four storied fluvial terrace-like spiral trench of 36-sided polygonal shape in several different magnitudes.

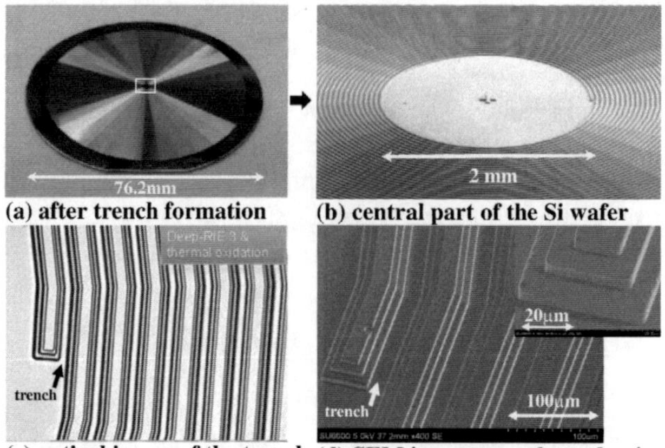

(a) after trench formation (b) central part of the Si wafer

(c) optical image of the trench including the end point of the spiral near the center (d) SEM image near the end point (the inset:a magnified image)of the trench near the center of the Si wafer

Fig.4 A spiral trench of a new 36-sided polygonal design fabricated on a Si wafer of 76.2 mm in diameter.

2.2. Deposition of superconducting NbN thin films succeeded by Cu electroplating and chemical mechanical polishing (CMP) process

After the formation of the spiral trench, reactive sputter-deposition of NbN was performed under the similar condition as those in our previous report [2]. Figures 5 (a) and (b) show how NbN film was deposited on each terrace and perpendicular wall in the trench. Although the overall cross-section of the trench is v-shaped but there still remain two sets of three perpendicular walls. Since sputtered particles are obliquely deposited onto these perpendicular walls of the trench, during the reactive sputtering of NbN, they are preferentially deposited on the salient portions of the periodic undulating pattern caused by repeated RIE (Bosh process), shadowing the portions of the perpendicular wall behind the salient portions. This self-shadowing effect caused inhomogeneous growth of NbN on the perpendicular walls again as in the previous study [3]. On the contrary, the homogeneous NbN thick film is formed on the bottom and terraces of the trench as thick as 3μm, thanks to the overall v-shaped cross-section of the newly designed trench. Fig.5(c) shows a whole image of the wafer after NbN deposition.

After the NbN deposition, the residual space in the trench was filled with Cu by electroplating. Fig.5 (d) shows a whole image of the wafer after the Cu plating. Since Cu has a high thermal conductivity, the Cu layer adjacent to the superconducting layer takes a roll to homogenize temperature fluctuation of the superconducting layer. This is essentially important to prevent accidental quenching of superconductivity when the superconducting current comes close to its critical current density. The second roll of the Cu layer is to bypass the current from the highly resistive superconducting layer in case of quenching of superconductivity to reduce the devastating effect of Joule heat. For these two rolls, the Cu layer must be in intimate contact with the superconducting layer. To attain this, deposition of NiP seeding layer was pre-deposited on the superconducting layer in the previous study [2]. However, the ferromagnetism of Ni may affect the superconductivity in the adjacent NbN layer. To help this, a thin Ti buffer layer was inserted between the NiP adhesive layer and the NbN layer to prevent direct magnetic exchange interaction by atomic layer deposition (ALD) utilizing its high throwing power. However, Cu seeding layer by non-electrolytic plating was tested in this study instead of NiP seeding layer to get rid of any concern about the ferromagnetism of Ni. Consequently, ALD of Ti buffer layer was also not used in this study. Fig.5 (e) shows a SEM image of a part of the cross-section of the trench. It is found that intimate contact between NbN layer and Cu layer was realized. Since the growth of Cu plating layer is most rapid at the edge of the entrance of the trench, the Cu plating layer is apt to block up the entrance, remaining the void space in the trench. To eliminate this, some inhibitors of Cu plating were also used depending on the location around the trench as in the previous study [2].

Then, the surface of the Si wafer is polished by CMP employing slurry of specially designed abrasives to eliminate the NbN layer and the Cu layer so that NbN and Cu remain only in the trench. In reality, CMP went beyond the original surface of the silicon wafer, so that the height of the first perpendicular wall of the trench decreased to 1.7-2.6 mm from the initial height of 8μm as shown in Fig.2. Fig.5 (f) shows a whole image of the wafer after CMP. Figs. 5(g) and (h) show optical microscope images around the end point of the spiral trench near the center of the wafer. Since the CMP went beyond the original surface of the Si wafer to the first terrace of the trench in this sample for observation, the NbN film grown on the first terrace is also visible. The dark portion is the Si bank and bright portion is the surface of Cu layer. The total length of the resultant traversable 3μm thick NbN film superconducting coil lined with Cu plated layer was 42m. The outer diameter of the spiral coil was 62 mm and the inductance was calculated to be around 3.8mH.

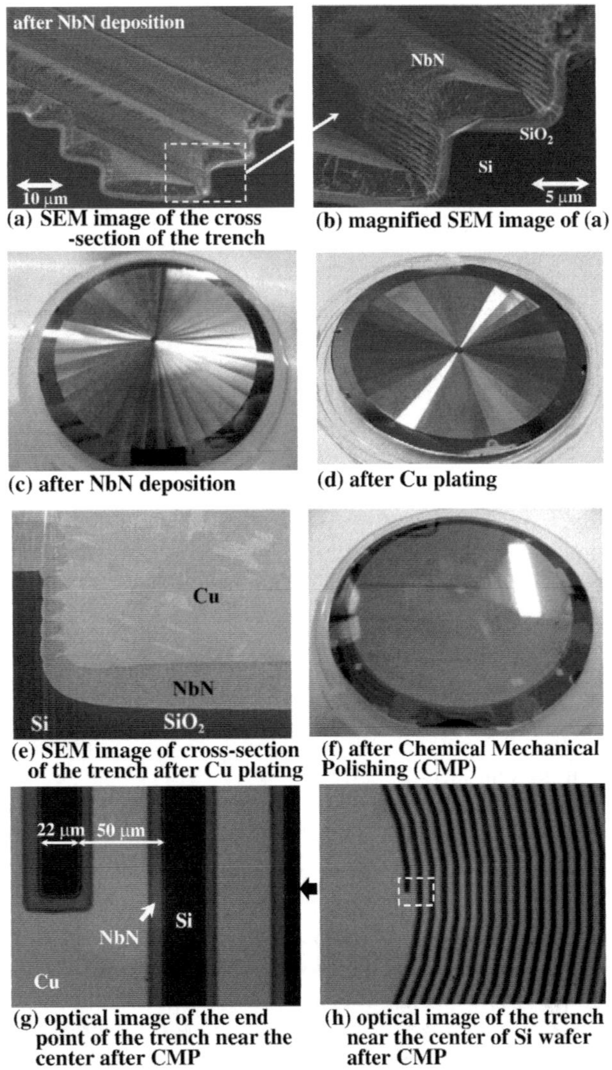

Fig.5 Photographs showing processes after the trench formation. (a)(b)(c)after NbN film deposition, (d)(e) after Cu plating, (f)(g)(h) after CMP

2.3. Measurement of superconducting properties

Superconducting properties of the one wafer sample shown in Fig.5(f) was measured using the same 4 K cryogenic system used for the measurement in the previous study. The one wafer sample was placed on the cold head (cryogenic sample stage panel) and connected with two probe wires. The two probe wires were attached at the Cu plated center field and the Cu plated outer edge field near the two ends of the spiral trench using silver paste. The handmade silver paste contacts are roughly 1mm in diameter.

Figures 6 (a) and (b) show the results of the measurements. Fig.6 (a) shows the temperature dependence of the resistivity between the two probes. With decrease in temperature, the DC resistivity showed a metallic decrease indicating the current pass was in the Cu plated layer and then made a sudden fall to residual contact resistance indicating the shift of current pass from the Cu plated layer to the NbN film at the critical temperature T_c of 15.5K by superconducting transition. Fig.6 (b) shows variations of current I flowing in the coil as a function of the applied voltage V between the two probes at different temperatures between 4K and 15K. In general, I increased with V in the initial stage along with a straight line defined by a contact resistance, indicating zero resistance in the coil. At a certain critical current I_c, the increase of I with V was detached from the straight line indicating that a finite resistivity arose in the coil. I decreased with V after having its maximum. With further increase in V, the variation of I approached asymptotically to a straight line indicating I-V relation caused by the resistivity of Cu plated layer coil. Therefore the critical current I_c is considered to be the superconducting critical current in the coil. *Ic* increased with decrease in temperature. The highest I_c in Fig.6 (b) was 220 mA at 4K. The amount of stored magnetic energy at this highest I_c was calculated to be 93 μJ and the corresponding energy storage volume-density was as high as 9.7Wh/*l*.

3. Discussion

In our previous study, the highest I_c was 43 mA at 3.7K and the amount of magnetic storage was 9.9 μJ in which the perfect superconducting coil on a Si wafer of 101.5 mm in diameter has not been successfully fabricated [2]. In the present study, the highest I_c attained using a smaller Si wafer of 76.2 mm in diameter was five times as large as that in the previous study and the amount of magnetic storage attained was almost 10 times as large as that in the previous study. The energy storage volume density of 9.7 Wh/*l* is fairly large in comparison with a typical value of conventional gigantic SMESs of 0.88 Wh/*l* [4].

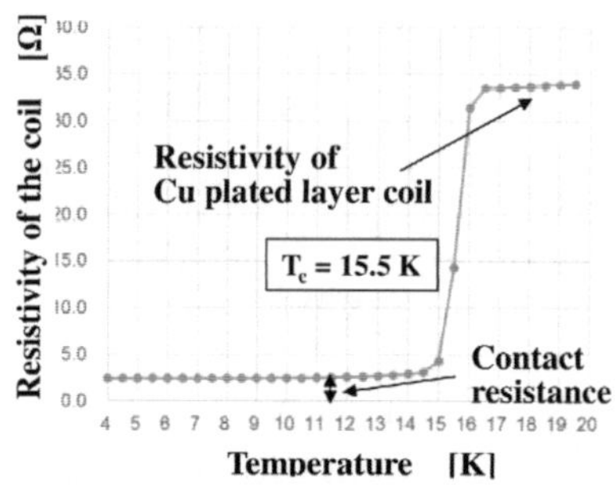

(a) Temperature dependence of the resistivity of the total coil between the center and the outer edge

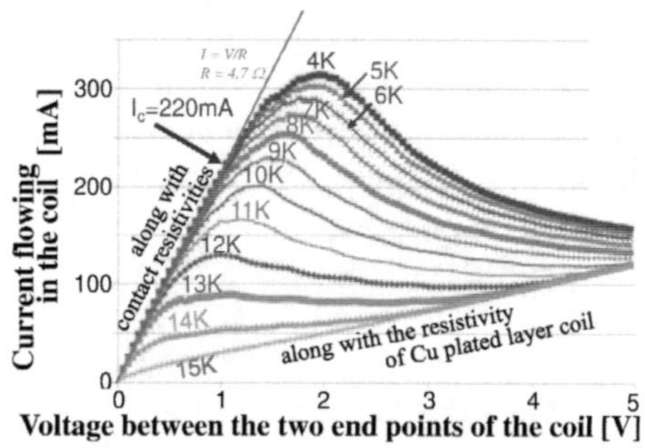

(b) Current voltage characteristic of the total coil between the center and the outer edge

Fig. 6 Measured superconducting properties of the total coil between the center and outer edge of the Si wafer of 76.2 mm in diameter

4. Conclusion

Based on the concept of a novel approach to make a compact SMES unit composed of a stack of Si wafers using MEMS process proposed previously [2], a complete fabrication of a traversable 3μm thick NbN film superconducting coil lined with Cu plated layer of 42m in length in a spiral three-storied trench engraved in and extended over a whole Si-wafer of 76.2 mm in diameter was attained for the first time. With decrease in temperature, the DC resistivity showed a metallic decrease indicating the current pass was in the Cu plated layer and then made a sudden fall to residual contact resistance of indicating the shift of current pass from the Cu plated layer to the NbN film at the critical temperature T_c of 15.5K by superconducting transition. The temperature dependence of *I-V* curve showed the increase in the critical current with decrease in the temperature and the highest critical current measured was 220 mA at 4K which is five times as large as that obtained in one third of the coil on a Si wafer of 101.5 mm in diameter in the test fabrication as the experimental proof of concept presented in the previous report. This completion of a one wafer superconducting NbN coil is an indispensable step for the next proof of concept of fabrication of series-connected two wafer coils via superconductive joint which will read to series connected 600 wafer coils finally, and for replacement of NbN by high T_c superconductor such as $YBa_2Cu_3O_{7-x}$ for operation under the cold energy of liquid hydrogen or liquid nitrogen.

Acknowledgement

This study is based on results from a project commissioned by NEDO (15101787-0, 15101788-0, 15101789-0, 15101790-0, 15101791-0).

References

[1] Ishida N, Iwasaki T, Sato R, Sato Y, Miyagi D, Tsuda M, Hamajima T, Shintomi T, Makida Y, Takao T, Iwaki K, 2014, *Oral Session: Renewable Energy and Hydrogen, Areal VII: Hydrogen & Fuel Cell, July 30, Grand Renewable Energy July 27-Aug. 1,* O-Hf-2-3, *Tokyo*, Japan.
[2] Sugimoto N, Iguchi N, Kusano Y, Fukano T, Hioki T, Ichiki A,Bessho T and Motohiro T, 2016, *Supercond. Sci. Technol.* 30**,** 015014
[3] Sugimoto N and Motohiro T 2012 *Vacuum* 93, p.13
[4] Nomura S, Shintomi T , Akita S, Nitta T, Shimada R and Meguro S 2010 *IEEE Trans. on Appl. Superconductivity* 20 p.1373

IOP Conf. Series: Journal of Physics: Conf. Series **897** (2017) 012020 doi:10.1088/1742-6596/897/1/012020

Removal of Iron Oxide Scale from Feed-water in Thermal Power Plant by Using Magnetic Separation

Motohiro Nakanishi[1], Saori Shibatani[1], Fumihito Mishima[2], Yoko Akiyama[1], Shigehiro Nishijima[1]

[1] Division of Sustainable Energy and Environmental Engineering, Graduate School of Engineering, Osaka University, A1 Building 2-1, Yamadaoka, Suita City, Osaka, 565-0871, Japan
[2] Department of Applied Nuclear Tecnology, Fukui University of Tecnology, 3-6-1, Gakuen, Fukui City, Fukui, 910-8505, Japan

E-mail: nakanishi@qb.see.eng.osaka-u.ac.jp

Abstract. One of the factors of deterioration in thermal power generation efficiency is adhesion of the scale to inner wall in feed-water system. Though thermal power plants have employed All Volatile Treatment (AVT) or Oxygen Treatment (OT) to prevent scale formation, these treatments cannot prevent it completely. In order to remove iron oxide scale, we proposed magnetic separation system using solenoidal superconducting magnet. Magnetic separation efficiency is influenced by component and morphology of scale which changes their property depending on the type of water treatment and temperature. In this study, we estimated component and morphology of iron oxide scale at each equipment in the feed-water system by analyzing simulated scale generated in the pressure vessel at 320 K to 550 K. Based on the results, we considered installation sites of the magnetic separation system.

1. Introduction

Recently, reduction of carbon dioxide (CO_2) has been required to prevent the global warming. It is necessary to keep the efficiency of thermal power plants high, because they account for the most of CO_2 emission accompanied with the electric power generation. One of the factors of deterioration in the efficiency is the adhesion of the iron oxide scale to the inner wall of pipe in feed-water system. [1] Iron oxide scale is main corrosion product with thermal conductivity as low as 10 % of pipework material. Therefore, the scale adhesion to inner wall of pipework decreases heat exchange efficiency. It is reported that the decline of thermal power generation efficiency caused by scale adhesion is 25 kW/μm in the 200 MW grade coal thermal power generation. [2] Thermal power plants have employed All Volatile Treatment (AVT) or Oxygen Treatment (OT) to prevent scale formation, though these treatments cannot prevent it perfectly. The effective method for removal of the scale has been required.

We proposed a high gradient magnetic separation (HGMS) system [3, 4] that can remove iron oxide scale without stopping thermal power plant. HGMS is a method to separate substances from fluid utilizing magnetic force. Ferromagnetic filters are installed into the flow channel and magnetized by external magnetic field. It is necessary for this system to treat 400~500 m^3/h feed-water, and hence we employed HGMS using superconducting magnet that can generate high magnetic field in wide area. Magnetic separation efficiency is largely influenced by component and morphology of scale. Scale changes their property depending on the type of water treatment and temperature of feed-water. In this study, we estimated component and morphology of iron oxide scale at each equipment in the feed-water system by analyzing simulated scale generated at 320 K to 550 K in the pressure vessel. Based on the results, we considered installation sites of magnetic separation system.

1.1. Water treatment
Two kinds of treatments, All Volatile Treatment (AVT) and Oxygen Treatment (OT), are performed in thermal power plants to prevent scale formation. In AVT, ammonia as pH adjuster and hydrazine as deoxidizing agent are added to feed-water. Corrosion of piping metal is reduced by keeping pH of feed-water high, and dissolved oxygen that induces corrosion is removed by hydrazine. On the other hand, ammonia and oxygen are added to feed-water in OT. Ammonia reduces corrosion of piping and oxygen changes the iron component into hematite (α-Fe_2O_3) that has lower solubility compared with piping materials. Passive film of hematite formed on inner wall protects piping material from corrosion. In this study, we subjected thermal power plant using AVT and OT since component and morphology of scale are different depending on the type of water treatment.

1.2. Mechanism of iron oxide scale formation
Figure 1 shows schematic diagram of thermal power plant. Feed-water treated by AVT or OT is kept pH high around 9, but the corrosion of pipe metal is promoted due to decrease in pH around 7 by degassing in the condenser. Then, pH of feed-water is raised by addition of ammonia, and eluted iron ions deposit as fine particles such as iron oxyhydroxide. Most of iron components change into iron oxide particles at high temperature area. [4] These particles deposit on the inner wall, and declines heat exchange efficiency.

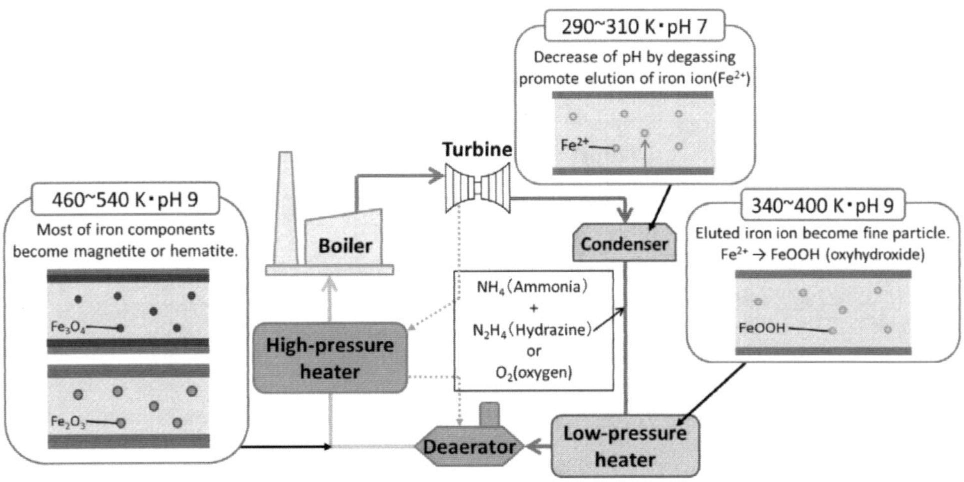

Figure 1. Schematic diagram of thermal power plant.

2. Experimental procedure
Figure 2 shows experimental procedure of simulated scale generation. We employed ferrous chloride solution and ferric chloride solution to simulate feed-water containing iron ion eluted from pipework. These solutions were prepared by dissolving iron (II) chloride tetrahydrate (KISHIDA CHEMICAL Co., Ltd.) and iron (III) chloride hexahydrate (KISHIDA CHEMICAL Co., Ltd.) to distilled water respectively, whose densities of iron ion were adjusted to 100 mg/L. When we simulated feed-water using AVT, ferrous chloride solution was used. The reason is that concentration of dissolved oxygen in feed-water using AVT is smaller than 10 μg/L and iron ion is hardly oxidized. 1 mol/L ammonia solution (NACALAI TESQUE, INC.) and hydrazine monohydrate (KISHIDA CHEMICAL Co., Ltd.)

were added to the solution. The pH of this solution and the concentration of hydrazine were adjusted to 9 and 10 mg/L respectively. On the other hand, ferric chloride solution was employed in simulation for OT. This is because feed-water using OT contains oxygen about 20~200 μg/L, so iron ion is well oxidized. Ammonia was added to ferric chloride solution to adjust pH to 9. These samples were enclosed into pressure vessel (Taiatsu Techno Corporation, TVS-N2) and heated for 24 hours. Heating condition is shown in table 1. Heating temperature was determined from temperature of feed-water at respective equipment of thermal power plant. Suspensions after heating were filtered and solid products were analyzed by XRD (Rigaku Corporation, Ultima IV) to identify the component of simulated scale. Particle size distributions of suspensions were measured before and after ultrasonic treatment to investigate morphology of simulated scale.

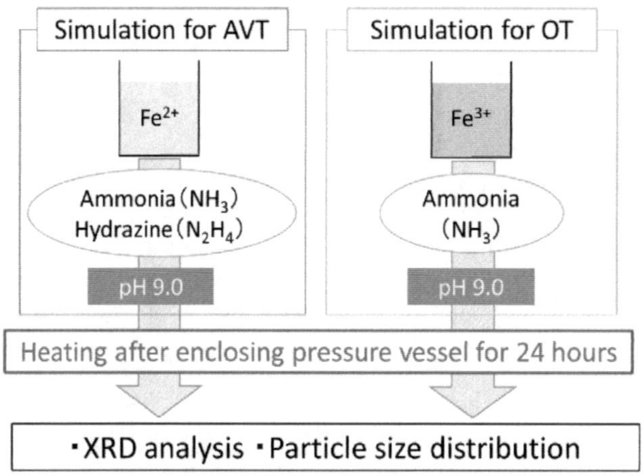

Figure 2. Experimental procedure of simulated scale generation experiment.

Table 1. Heating condition.

Equipment	Temperature[K]	Pressure[MPa]
Inlet of low pressure heater	320	0.10
Outlet of low pressure heater	400	0.27
Outlet of deaerator	450	1.0
Drain of high pressure heater	450	1.0
Outlet of high pressure heater	550	6.4

3. Results and Discussions

3.1. Simulation for AVT

Figure 3 shows XRD spectrums of generated scales simulating AVT thermal power plant. From this figure, the characteristic peak positions of simulated scales at every heating temperature were corresponded to that of magnetite. It indicates that main component of the scale is magnetite in AVT thermal power plant. Magnetite is ferromagnetic that can be removed by magnetic separation easily. Figure 4 shows spectrum half-width of simulated scales and commercially available magnetite (MITSUI MINING & SMELTING CO.,LTD.) at the peak of $2\theta = 35.4\ ^\circ$ in XRD spectrum. Figure 4 shows low crystallinity of simulated scales compared with commercial magnetite because low half-width represents high crystallinity. Low crystallinity means low magnetic property and hence we should consider difference of magnetic property between literature data and real scale's characteristic in developing magnetic separation system.

IOP Conf. Series: Journal of Physics: Conf. Series **897** (2017) 012020 doi:10.1088/1742-6596/897/1/012020

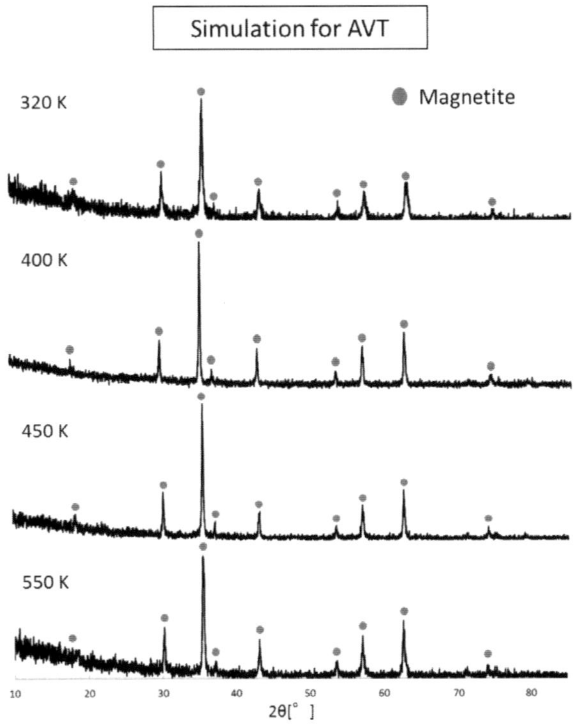

Figure 3. XRD spectrum of scale in the case of simulation for AVT.

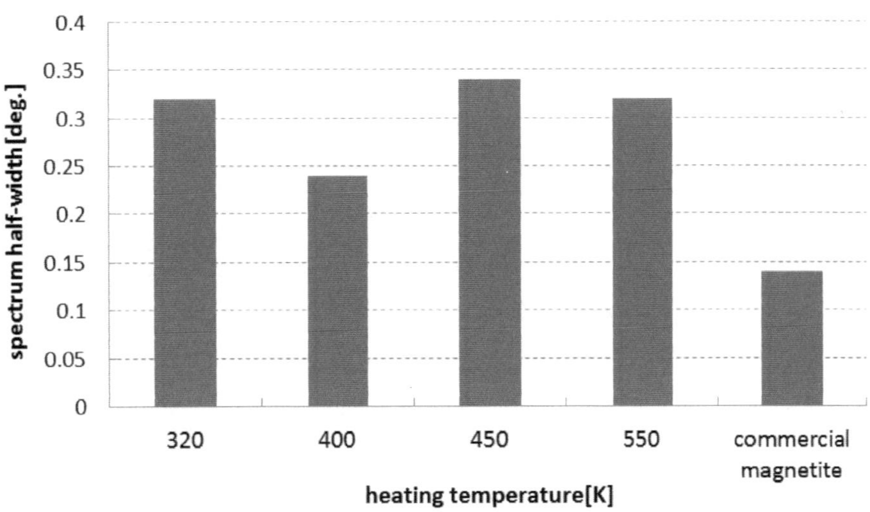

Figure 4. Spectrum half-width of simulated scale and commercial magnetite.

Figure 5 shows particle size distribution of generated scales simulating for AVT. These measurements were conducted before and after ultrasonic treatment. From this figure, particle size of simulated scale at all heating temperature became smaller by ultrasonic treatment. This result indicates aggregation of simulated scale. In practical condition, it is supposed that the scale is aggregated, and hence the dispersion due to flow of feed-water is also considered. Particle size of simulated scale heated at 320 K was smaller than that at other heating temperatures. It can be estimated that particle size of practical scale at low-temperature area such as inlet of low-pressure heater is smaller than that in high temperature area.

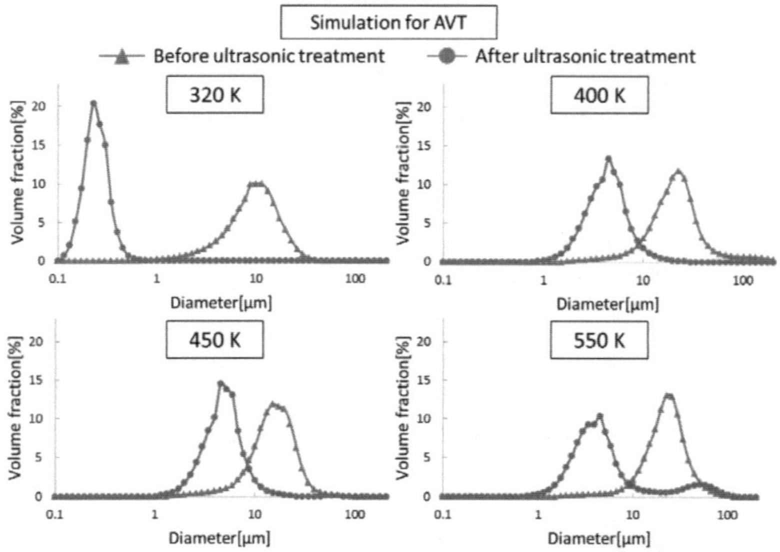

Figure 5. Particle size distribution of simulated scale.

From the above, main component and median size of scale in AVT thermal power plant estimated are shown in figure 6. In magnetic separation, larger particle size of target particle increases higher magnetic separation efficiency. Therefore, it is thought that magnetic separation system should be installed into high temperature area after low-pressure higher in AVT thermal power plant.

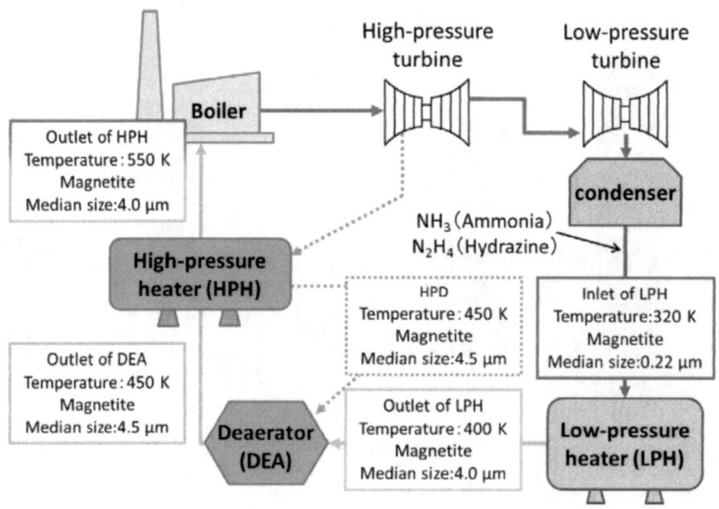

Figure 6. Component and median size of scale estimated from analysis (AVT).

3.2. Simulation for OT

Figure 7 shows XRD spectrums of generated scale simulating OT thermal power plant. XRD spectrum of simulated scale heated at 320 K had no peak, whereas peak positions of scales heated at 400, 450 and 550 K were corresponded to those of hematite. From this result, the scales in OT thermal power plant are noncrystalline at inlet of low-pressure heater and hematite at high temperature area. This noncrystalline substance is inferred iron oxyhydroxide which was generated when ammonia was added to solution containing iron ion [5]. Iron hydroxide and hematite are both paramagnetic and hence they can be removed by HGMS using superconducting magnet. The spectrum half-width of simulated scales and commercially available hematite (TODA KOGYO CORP.) at the peak of 2θ = 33.1 ° in XRD spectrum are shown in figure 8. Low crystallinity of simulated scales compared with commercial hematite was suggested by the figure. We should consider that crystallinity effects on the magnetic property in examining magnetic separation system as with the case of AVT.

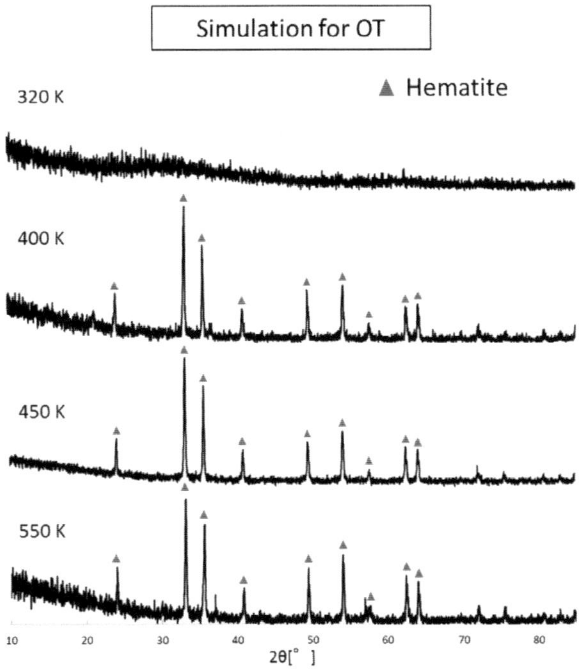

Figure 7. XRD spectrum of scale in the case of simulation for OT.

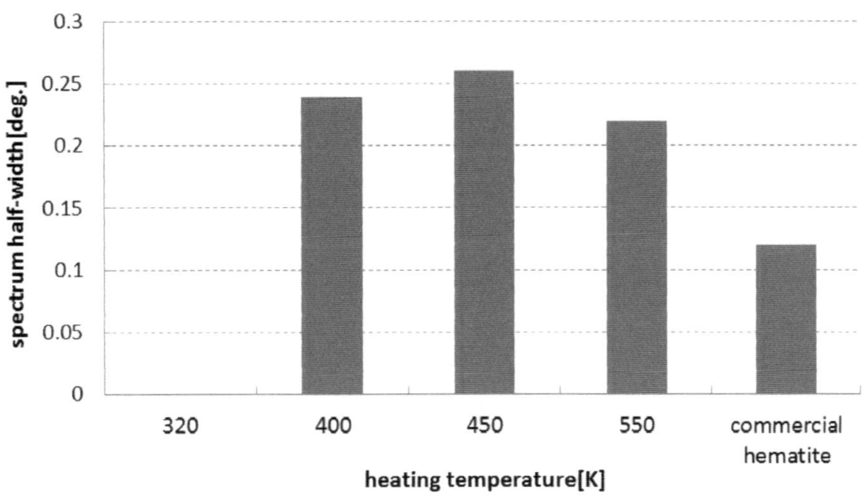

Figure 8. Spectrum half-width of simulated scale and commercial hematite.

Figure 9 shows particle size distribution of generated scales simulating for OT. These measurements were conducted before and after ultrasonic treatment. Figure 9 shows aggregation of simulated scale because particle size became smaller by ultrasonic treatment for all heating temperature. In practical condition, it is supposed that the scale is also aggregated, though dispersion due to flow of feed-water is considered. Particle size distributions of these simulated scale heated at 400, 450 and 550 K after ultrasonic treatment had broader peak in respect of those of generated scale simulating for AVT. Therefore, it is thought that the scale in OT thermal power plant has wider particle size distribution compared with that in AVT. It is reported that particle size of amorphous ferric oxyhydroxide is observed to be less than 8 nm by transmission electron microscope [5], and consequently the simulated scale heated at 320 K would not be dispersed by ultrasonic treatment completely.

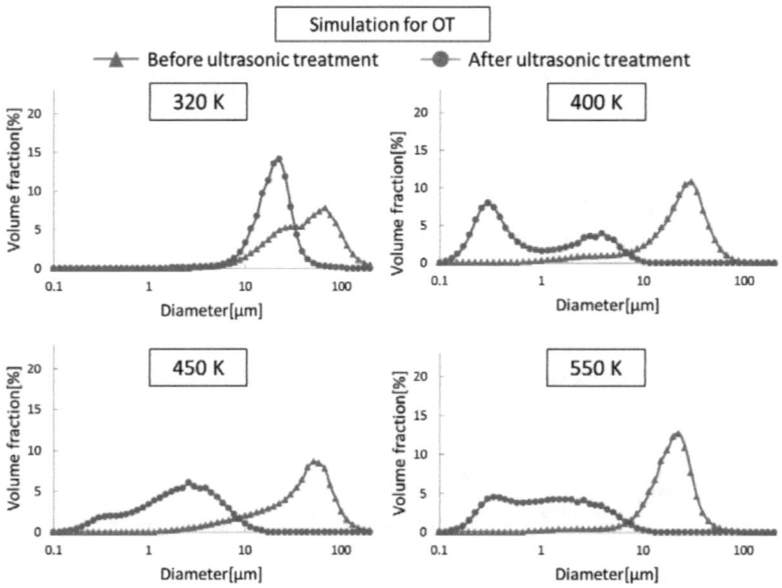

Figure 9. Particle size distribution of simulated scale.

Main component and median size of scale in OT thermal power plant estimated from the analysis are shown in figure 10. In magnetic separation, larger particle size of target particle results in higher magnetic separation efficiency. Therefore, it is considered that we should install magnetic separation system into outlet of deaerator (DEA) or drain of high-pressure heater (HPD) in OT thermal power plant.

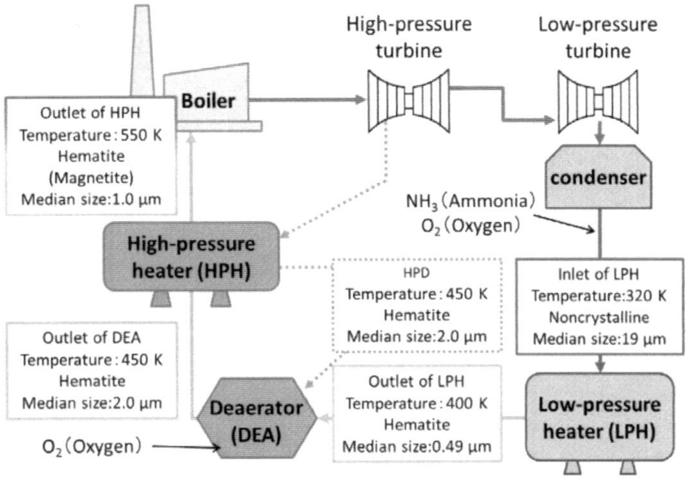

Figure 10. Component and median size of scale estimated from analysis (OT).

4. Conclusion

We proposed a HGMS system using superconducting magnet in order to remove iron oxide scale. In this study, simulated scales generated by using pressure vessel heated at 320~550 K were analyzed. Based on the results, we estimated component and morphology of practical scale generated in the AVT or OT feed-water of thermal power plant. Iron oxide scale can be removed efficiently by magnetic separation when we install this system at high temperature area in feed-water system of AVT thermal power plant, and outlet of deaerator or drain of high-pressure heater of OT thermal power plant. In practical condition, flow flux or density of iron oxide scale should be also taken into account for decision of installation site. Moreover, difference of magnetic property between commercial iron oxide and real scale products should be examined. In future, we are going to investigate magnetic separation efficiency with magnetic separation experiment for these simulated scales.

Acknowledgements

This research was partly supported by "Advanced Low Carbon Technology Research and Development Program (ALCA)" of Japan Science and Technology Agency (JST).

References

[1] Zhong X, Wu X and Han E, 2012, "The characteristic of oxide scales on T91 tube after long-term service in an ultra-supercritical coal power plant," *J. of Supercritical Fluids*, vol. **72**, pp. 68- 77.

[2] Siddhartha Bhatt M, 2006, "Effect of water side deposits on the energy performance of coal fired power plants,", *Energy Conversion and Management*, vol **47**, pp. 1247-1263.

[3] Nishijima S, Takahata K, Saito K, Okada T, Nakagawa S and Yoshiwa M, 1987, "Applicability of superconducting magnet to high gradient magnetic separator," *IEEE Trans. Magn.*, vol. **23**, no. 2, pp.573-576.

[4] Mizuno N, Mishima F, Akiyama Y, Okada H, Hirota H, Matsuura M, Maeda T, Shigemoto N and Nishijima S, 2015, "Removal of iron oxide with superconducting magnet high gradient magnetic separation from feed-water in thermal plant," *IEEE Trans. Appl. Supercond.*, vol. **25**, no. 3.

[5] Iwasaki K, Itoh T and Yamamura T, 1999, "Effect of pH of Colloidal Mother Liquid on Formation of Acicular α-FeOOH Particle Made from Amorphous Ferric Oxydydroxide," *Journal of the Japan Society of Powder and Powder Metallurgy*, vol **46**, no. 3.

Study on AC loss measurements of HTS power cable for standardizing

Shinichi Mukoyama[1], Naoyuki Amemiya[2], Kazuo Watanabe[3], Yasuhiro Iijima[3], Nobuhiro Mido[4], Takao Masuda[5], Toshiya Morimura[5], Masayoshi Oya[5], Tetsutaro Nakano[6], and Kiyoshi Yamamoto[7].

1 Furukawa Electric Co,, Ltd. Telecomunication & Energy Laboratories, 6, Yawatakaigan-dori, Ichihara, Chiba 290-8555, Japan
2 Kyoto University, Department of Electrical Engineering, Kyoto-Daigaku-Katsura, Nishikyo, Kyoto 615-8510, Japan
3 Fujikura, Energy Technology Research Dept., Advanced Technology Lab. 1440, mutsuzaki, sakura, chiba 285-8550, Japan
4 Showa Cable System Company Ltd., Superconductor Technology Center, 1-1, Minami Hashimoto 4-chome, chuo-ku, Sagamihara-city, Kanagawa, 252-0253 Japan
5 Sumitomo Electric, Superconductivity Technology Division, 1-1-3, Shimaya Konohana-ku, Osaka 554-0024, Japan
6 Tokyo Electric Power Company Holdings, TEPCO research institute, 4-1, Egasaki-cho, Tsurumi-ku, Yokohama 230-8510, Japan
7 The Japanese Electric Wire & Cable Makers' Association, Japan National Committee for the IEC/TC90 Superconductivity, Konwa-Bldg., 1-12-22, Tsukiji, Chuo-ku, Tokyo, 104-0045 Japan

(shinichi.mukoyama@furukawaelectric.com)

Abstract. High-temperature superconducting power cables (HTS cables) have been developed for more than 20 years. In addition of the cable developments, the test methods of the HTS cables have been discussed and proposed in many laboratories and companies. Recently the test methods of the HTS cables is required to standardize and to common in the world. CIGRE made the working group (B1-31) for the discussion of the test methods of the HTS cables as a power cable, and published the recommendation of the test method. Additionally, IEC TC20 submitted the New Work Item Proposal (NP) based on the recommendation of CIGRE this year, IEC TC20 and IEC TC90 started the standardization work on Testing of HTS AC cables. However, the individual test method that used to measure a performance of HTS cables hasn't been established as world's common methods. The AC loss is one of the most important properties to disseminate low loss and economical efficient HTS cables in the world. We regard to establish the method of the AC loss measurements in rational and in high accuracy. Japan is at a leading position in the AC loss study, because Japanese researchers have studied on the AC loss technically and scientifically, and also developed the effective technologies for the AC loss reduction. The JP domestic commission of TC90 made a working team to discussion the methods of the AC loss measurements for aiming an international standard finally. This paper reports about the AC loss measurement of two type of the HTS conductors, such as a HTS conductor without a HTS shield and a HTS conductor with a HTS shield. The AC loss measurement method is suggested by the electrical method..

1. Introduction

High temperature superconducting cables (HTS cables) use HTS wires, which perform zero resistivity and high current density, in a cable conductor. The HTS cable has a low transmission loss and a compact size compare to the copper or aluminium convention power cables. DC current can flow in a superconductor with zero resistance. However concerning with AC power cables, the superconductor generates AC loss due to a magnetic-flux flow by an electrical magnetic force generated by an AC current flowing. Moreover a coupling loss and an eddy current loss induced in metal matrix of HTS wire add the AC loss of the cable. It is important to reduce the AC loss for realizing economical advantage and energy-saving of the HTS cable. AC loss reduction is most important and recently the low AC loss of 1 W/m is achieved in the 3 kA HTS cable.

Since methods of the AC loss measurement have been proposed by many laboratories, there is not the standard method that all researchers recognized. A lot of AC loss measurement methods have been conducted and proposed in the all over the world. Recently AC loss in the operated condition is important, because the value of the AC loss decide a capacity of the cooling system, and estimate the performance of the HTS cable to customers. Therefore, to commercialize the HTS cable, the AC loss measurement is required not only in objectivity and accuracy but also in easily for the manufacturers.

Researchers and engineers in Japanese domestic committee of IEC 90 have discussed about the method of the AC loss measurement to standardize it under supporting by Ministry of Economy, Trade and Industry

2. Survey of AC loss measurement

The AC loss measurements have been conducted by a lot of laboratories around the world. There are two methods to measure the AC loss. One is the electrical method, which is called an AC four terminal method[1[-[5], and the other is the calorimetric method[6]-[9], in which we measure the heat generated from the AC loss.

2.1. Electrical method

A typical electrical method of the AC loss measurement is the procedure of putting voltage taps on the both ends of the cable conductor and of measuring the cable voltage during loading current. Since the cable voltage consists of the voltage from inductance (L component) and from resistance (R component), it is necessary to omit the large amount of voltage from L component. The two methods of omitting the L component are proposed. One is to use a lock in amplifier, which measures the same phase voltage as the phase of the current. The other is to use a data recorder, which measures the voltage and the current at short time interval, and to summarize multiplications of the voltage and the current for one period with a computer. In some cases, to raise accuracy of the measurement the L component is cancelled by using the pick-up current. Many studies about how to arrange the return current pass and how to conform the voltage tap and the voltage lead were conducted for accuracy improvement of the measurement. A suggestion of AC loss of a conductor with a shield was also studied. The electrical methods provide AC loss measurement with measurement range from 10^{-14} W/m to 10 W.

2.2. Calorimetric method

The calorimetric method measures a minute calorie generated by AC loss in the cable conductor. Procedures of the measurement of the heat flux were explained three methods that, such as measuring of temperature rising of the conductor, measuring of boil off of the evaporated nitrogen gas, and measuring of temperature difference between inlet nitrogen and outlet nitrogen. In all methods, it is important to remove inversion heat from the cryostat pipe, joule loss in electrode, and conducting heat from cable ends.

3. Recommendation of AC loss measurement

An industrial standard needs not only reliability and accuracy but also simple for the measurement, because cost of the measurement adds price of a HTS cable. The high cost prevents to spread the HTS cables to the market, and doesn't provide benefit to the manufacturers and the users. We propose the electrical method as the simple and easy method. We describe the procedure of the measurement in following.

3.1. AC loss measurement of the conductor without HTS shield

An open type bath that has enough volume for entering a test sample is filled with liquid nitrogen, and a HTS cable sample is immersed in it. Electrodes that use for applying current are formed on the both end of the HTS cable sample. The voltage taps are put on the surface of the conductor at inside of the electrodes. The electrodes are connected to an AC current source with insulated Cu cables. AC current applied to the sample is measured using a current detector such as shunt-resistor, a current-transformer, a Rogowski-coil, etc. The voltage lead wires connected on the voltage taps generally are disposed on the surface of the conductor, and two lead wires are twisted at the center of the conductor and twisted wire is connected to a voltage measurement device.

The AC loss is calculated by a time integration of current and voltage over one period. The current and the voltage are measured and are recorded with a data acquisition device at short time interval, the both recorded values of current and voltage are multiplied, and the multiplied values are numerical integrated for one period, for example using a computer. It is shown as the second term of equation 1. To improve sensitivity of resistive component of the voltage, some laboratories canceled the inductive component of the voltage by using current phase voltage from a coupling coil in the current loop.

When the current and the voltage are sine-wave with little deformation, the AC loss is induced with a lock-in-amplifier. The lock-in-amplifier can directly explain the phase difference (θ) between the current and the voltage, and the AC loss is calculated by the third term of the equation 1.

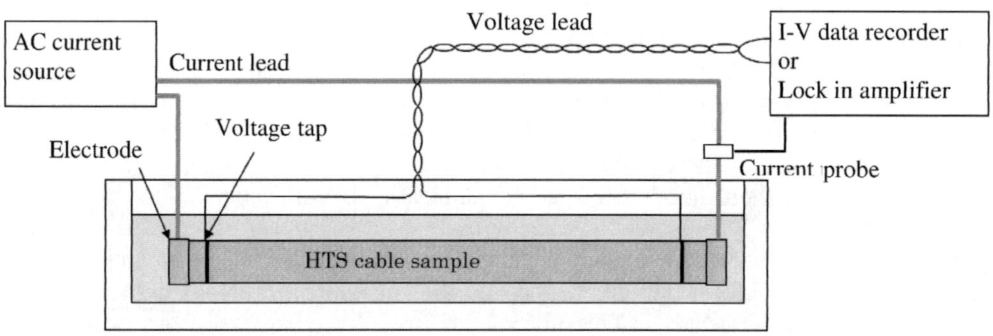

Figure 1. Diagram of AC loss measurement. Measurement device of AC loss is used I-V a data recorder or a lock-in-amplifier

$$P = \frac{1}{T}\int_0^T I(t) \times V(t)dt = \sum I(t) \times V(t) = I_e \times V_e \cos\theta \qquad (1)$$

,where $I(t)$ is current in the conductor, $V(t)$ is voltage between voltage taps, T is time of one period, θ is phase difference between I(t) and V(t), I_e is effective value of the current, and V_e is effective value of the voltage.

IOP Conf. Series: Journal of Physics: Conf. Series **897** (2017) 012021 doi:10.1088/1742-6596/897/1/012021

3.2. AC loss measurement of the conductor with HTS shield

In the case of the HTS cable sample with a HTS shield layer, we propose the way of AC current flows in the conductor layer and the shield layer in a series by jointed the conductor and the shield at the one end with a Cu lead. In a long-distance HTS cable, actually AC current that flows in the shield layer is almost same volume with counter direction against the conductor current, because the shield current is induced from the magnetic field produced by the conductor layer and the shield layer of the long-distance cable has extremely small resistance in a cable operation current compared to the inductance of the shield layer. Therefore the current flowing in series between the conductor layer and the shield layer can be adopted to measure the AC loss for the standardization, when the commercial HTS cable will is estimated in the operation condition of the real grid.

We are studying three ways of the AC loss measurement in the cable with shield layer. The voltage taps that measure cable voltages are formed at each end of the conductor and the shield, the AC loss is derived from the current and the voltage measured with the voltage tapes. Methods of (2) and (3) are developed and examined its verification of by Sumitomo Electric.

(1) The end to end voltages of the conductor and the shield are measured. The each AC loss is derived from the voltage and the current in the conductor and the shield respectively, by using a data integrated device or a lock-in-amplifier. (Each Layer Individual Measurement Method)

(2) Total AC loss of a whole system consists of a conductor and a shield is derived using voltage taps on the other side from the jointed end. The AC loss of the HTS cable is the value calculated by subtraction total AC loss of the system and AC loss of the jointed part. (Joint Part Subtraction Method)

(3) The voltage between the one side voltage tap on the conductor and the one side voltage tap on the shield and the voltage of the joint part in the other side are inputted in a lock-in-amplifier with differential mode (A-B). The AC loss is derived from the voltage of the differential signal and the phase difference (θ) against the AC current. (Joint Part Differential Method)

1) Each Layer Individual Measurement Method

Figure 2 shows the measurement diagram for measuring of the AC loss in the conductor and the shield respectively. V_{AB} of the end to end voltage on the conductor and V_{CD} of the end to end voltage on the shield are measured with a data recorder or a lock-in amplifier. The AC loss was derived from resistive component of the current and the voltage as equation 2.

$$Q = \int_0^T (V_{AB} \times I)dt + \int_0^T (V_{CD} \times I)\, dt = I \times V_{AB} \cos\theta_C + I \times V_{CD} \cos\theta_S \quad (2)$$

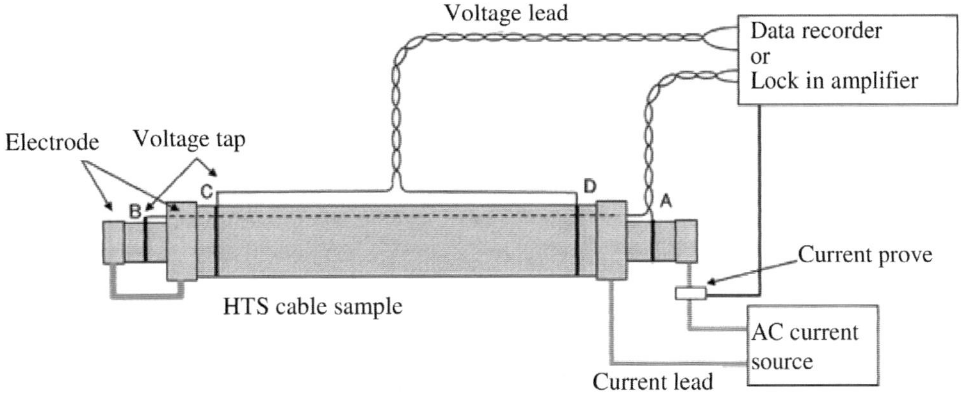

Figure 2. Diagram of AC loss measurement (Each Layer Individual Measurement Method)

2) Connection Part Subtraction Method
Figure 3 shows the measurement diagram of AC loss measurement that derives from the total AC loss subtracted the AC loss of the joint part. The AC loss is derived from V_{AD} of the end to end voltage between point A and point D and V_{CD} of the end to end voltage between point B and point C measured with a data recorder or a lock-in amplifier. The AC loss was calculated by equation 3.

$$Q = \int_0^T (V_{AD} \times I)dt - \int_0^T (V_{BC} \times I)\, dt = I \times V_{AD}\cos\theta_{AD} - I \times V_{BC}\cos\theta_{BC} \tag{3}$$

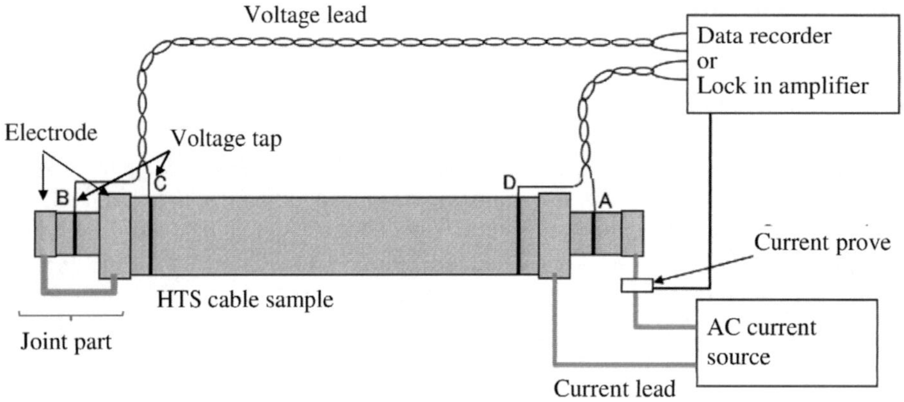

Figure 3. Diagram of AC loss measurement (Joint Part Subtraction Method)

3) Joint Part Differential Method
Figure 4 shows the measurement diagram of AC loss measurement by a lock-in-amplifier, which operate differential mode of input A and input B. V_{AB} of the end to end voltage is inputted to terminal A of the lock-in-amplifier and V_{CD} of the end to end voltage is inputted to terminal B. The AC loss of the cable is measured from the voltage of (A-B), the current, and the phase difference by equation 4.

$$Q = I \times (V_{AD} - V_{BC})\cos\theta \tag{4}$$

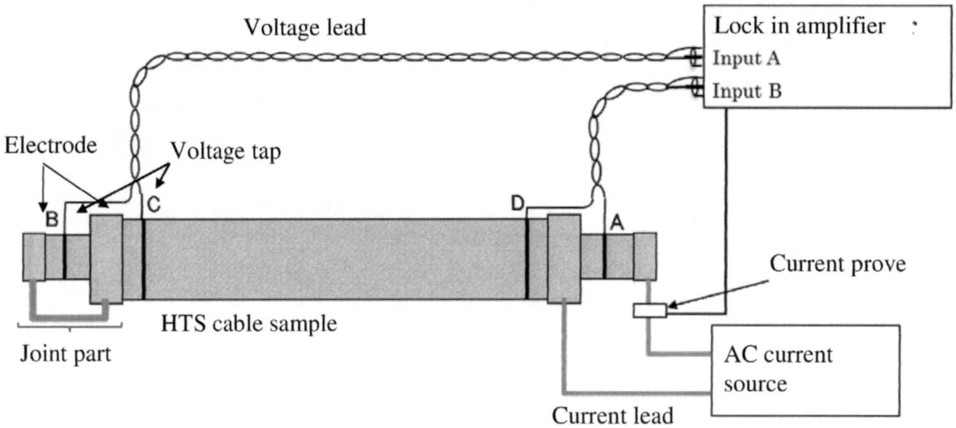

Figure 4. Diagram of AC loss measurement (Joint Part Differential Method).

IOP Conf. Series: Journal of Physics: Conf. Series **897** (2017) 012021 doi:10.1088/1742-6596/897/1/012021

4. Principal of electrical measurement.

4.1. Poynting vector

The electrical method (AC four-terminal method) is being proposed as AC loss measurement. We examine a validity of the AC four-terminal method for the AC loss measurement theoretically.

One of evaluation methods of AC loss generated in superconductor is a calculation method that integrate Poynting vector (E x H) on closed surface. The Poynting vector shows flow of electro-magnetic energy. Divergence of vector product $(E\times H)$, where E is electrical field and H is magnetic field, is developed as equation (5);

$$\nabla\cdot(\mathbf{E}\times\mathbf{H})=\mathbf{H}\cdot\nabla\times\mathbf{E}-\mathbf{E}\cdot\nabla\times\mathbf{H} \tag{5}$$

Rotation of electrical field and magnetic field in the right term is modified by Maxwell equations, and the both terms are arranged after volume integral in arbitrary closed volume V. The result becomes equation 6.

$$-\int_S (E\times H)dS=\int_V E\cdot J dV+\int_V E\cdot\frac{\partial D}{\partial t}dV+\int_V H\cdot\frac{\partial B}{\partial t}dV \tag{6}$$

, where *J, D, and B* are respectively density of current, density of electrical flux, and density of magunetic flux. These values has relationship by equation 7, and equation 8, using permittivity ε and permeability μ,

$$D=\varepsilon E \tag{7}$$

$$B=\mu H \tag{8}$$

The AC loss is obtained from equation 6, where the Poynting vector is surface-integrated on the closed area (S) enclosed the volume (V) and furthermore time-integrated for one period of AC. In the case of the volume doesn't have ferromagnetic material and ferroelectric material, equation 6 becomes equation 9. The AC loss is derived from the Poynting vector.

$$\mathrm{Q}=-\int_T dt\int_S (E\times H)dS=-\int_T dt\int_V E\cdot J dv \tag{9}$$

4.2. AC loss measurement of conductor without shield layer

HTS cable consisted of only conductor without shield is simulated as figure 5, which shows a cylindrical conductor of radius (R_C).

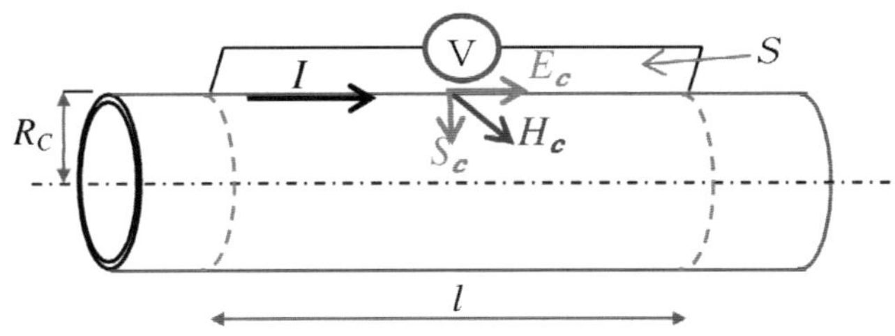

Figure 5. AC loss analysing model of a HTS conductor without a shield layer.

The magnetic field (H_C) and the electrical field (E_C) are provided from equation 8 in uniform magnetic field and electrical field along circumferential direction and axial direction. Since some actual HTS cables consist of HTS tapes wound spirally around a former, there is longitudinal magnetic field along axial direction. We neglected the longitudinal field in the basic analysis, because we simplified the model. This model is realized by non-winding tape on the conductor.

$$H_C = \frac{I}{2\pi R_C}, \quad E_C = \frac{V_C}{l} \tag{8}$$

, where I is cable current, Vc is voltage between voltage taps, and l is distance between the taps. The equation 7 that is substituted equation 8 for is surface-integrated on S-surface. The AC loss per unit length is shown as equation 9, and equation 9 shows that the AC loss is derived by using the current and the voltage in the conductor.

$$Q = \frac{1}{l}\int_0^T dt \frac{I}{2\pi R_C}\frac{V_C}{l}\int_S ds = \frac{1}{l}\int_0^T dt \frac{I}{2\pi R_C}\frac{V_C}{l} 2\pi l = \frac{1}{l}\int_0^T V_C I dt \tag{9}$$

4.3. AC loss measurement of conductor with shield layer

Divergence of Poynting vector on a shield layer is considered as in and out of the Poynting vector from a coaxial cylinder surface enclosed by surface of inside of shield and surface of outside of the shield. Counter direction current against the conductor current flows in the shield in the condition of Figure. 2, the magnetic field in outside of the shield is zero. The Poynting vector in the outside is zero. The AC loss from the Poynting vector inside the shield is indicated by equation 10 as figure 6.

$$Q = \frac{1}{l}\int_T dt \int_S E_S \cdot H_S ds \tag{10}$$

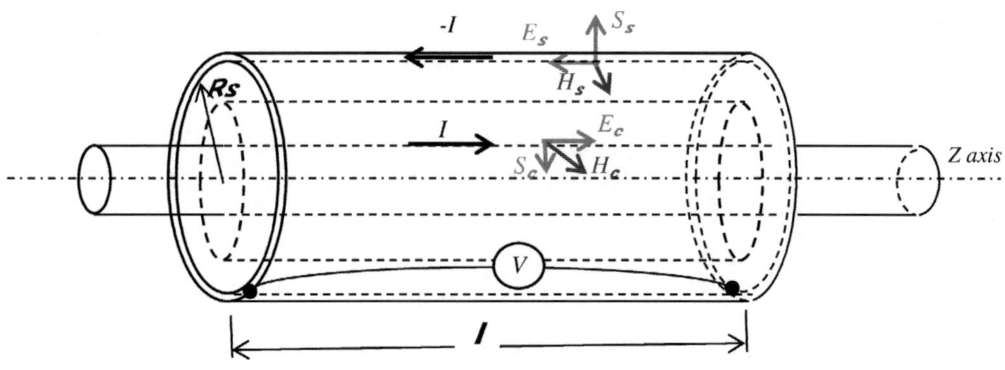

Figure. 6 AC loss analysing model of a HTS conductor with a shield layer.

The electrical field and the magnetic field is provided from equation 11 in uniform along circumferential direction and axial direction like the conductor examination and shield current in the closed surface enclosed by the shield inside surface. We neglected it the same as the conductor without shield layer in section 4.2.

$$H_S = \frac{I}{2\pi R_S}, \quad E_S = \frac{V_S}{l} \tag{11}$$

$$Q = \frac{1}{l}\int_0^T dt \frac{I}{2\pi R_S}\frac{V_S}{l}\int_S ds = \frac{1}{l}\int_0^T V_S I dt \tag{12}$$

The AC loss per unit length of the shield layer is derived by using the current and the voltage in the conductor as same as the conductor. In the case of actual shield layer, the voltage taps is not put on inside surface of the shield in the structure. The voltages of the inside surface and the outside surface are same, because multi-layer shields is gather at cable end with solder and any cross section of shield has the same potential of voltage along the axis.

5. Summary

AC loss is important parameter for realization of HTS cables, and many researchers have been developing AC reduction technologies for long time. AC loss measurement that can be used in common wide currency is required, and the international standard will be requested for the merchandize. A lot of experiences and data is needed for establish of the standard. The proposal of the AC loss measurement in the paper has two objectives; one is that all researchers will test by the same method for accumulating data, and the other is that discussion about AC loss measurements will start. In this progress, the standard will be established under international agreements.

6. Acknowledgement

This work is supported by Ministry of Economy, Trade and Industry. Authors also appreciate Dr. Teruo Matsushita with Kyushu Institute of Technology.

Reference

[1] Jiahui Zhu, et.al., "Electric Measurement of the Critical Current, AC Loss, and Current Distribution of a Prototype HTS Cable", *IEEE Transactions on Applied Superconductivity*, Vol. 24, No 3 (2014)

[2] Li, Z.Y., et.al., "Loss Test for a 5-m YBCO Cable Sample of the 22.9-kV KEPCO System Under Grid Operation" , *IEEE Transactions on Applied Superconductivity*, Vol. 23, No 3 (2013)

[3] Nguyen, D.A, et.al., "Measurements of AC Losses and Current Distribution in Superconducting Cables", *IEEE Transactions on Applied Superconductivity*, Vol. 21, No.3 (2011)

[4] Vysotsky, V.S, et.al., "AC Loss and Other Researches with 5 m HTS Model Cables", *IEEE Transactions on Applied Superconductivity*, Vol. 21, No.3 (2011)

[5] J. Souc, et.al., "AC Loss Measurement of YBCO Cable Model", *IEEE Transactions on Applied Superconductivity*, Vol.17, No. 2 (2007)

[6] D.E.Daney, et.al., "Calorimeter for measuring AC losses in HTS cables for superconducting power transmission lines", *Cryogenics* 39 (1999)

[7] B ten Haken, et.al., "Advanced test beds for quality control of superconductors in power applications", *Supercond.Sci.Technol.*13(2000)

[8] C.Træholt, et.al., "Calorimetric Measurements of Losses in HTS Cables", *IEEE Appl. Supercond. Proceedings* (2011)

[9] M. Yagi, et.al., "Measurement of AC Losses of Superconducting. Cable by Calorimetric Method and Development of. HTS Conductor With Low AC Losses" *IEEE Transactions on Applied Superconductivity*, Vol. 13, No. 2, (2003)

IOP Conf. Series: Journal of Physics: Conf. Series **897** (2017) 012022 doi:10.1088/1742-6596/897/1/012022

Design and Performance Test of Superconducting Transport Solenoid for D-Line at J-PARC Muon Science Facility

Yasuyuki Tanaka[1,#], Tomoyuki Semba[1], Shotaro Nakajima[1], Yoshiaki Hagiwara[1], Shuichi Kido[1], Yukihiro Murata[2], Ken-ichi Sasaki[3], Koichiro Shimomura[3], Naritoshi Kawamura[3], Patrick Strasser[3], Yasuhiro Makida[3], Hirokatsu Ohhata[3], Noriyuki Kurosawa[3], Yasuhiro Miyake[3]

[1]Hitachi, Ltd. Hitachi Works,
3-1-1, Saiwai-cho, Hitachi, Ibaraki, 317-8511, Japan
[2] Hitachi, Ltd. Research and Development Group,
7-1-1 Omika-cho, Hitachi, Ibaraki, 319-1292, Japan
[3] High Energy Accelerator Research Organization (KEK),
1-1, Oho, Tsukuba, Ibaraki, 305-0801, Japan

[#] yasuyuki.tanaka.xj@hitachi.com

Abstract. A superconducting transport solenoid for Decay Muon Line (D-line) at J-PARC Muon Science Facility was newly designed and manufactured. It was designed to generate a magnetic field in relatively large region (warm bore diameter 0.2 m), while keeping the same outer dimensions, connection interfaces to the existing refrigerator and the power supply of the previous machine [1-3]. Major changes of both solenoids are the reduction of the central magnetic field, the equipment of a warm bore and the adoption of the high T_c current leads. After the installation to the beam line, the initial cooling test, the excitation test and the emergency shutdown test at the rated current were conducted by KEK in order to confirm cryogenic and magnetic performance. These tests were successfully performed with no damege and indicated the solenoid was precisely manufactured and fulfilled the requirements. The solenoid has been under operation since July, 2015. This report describes the design, the manufacturing process, the magnetic field measurement at room temperature and the results of performance tests conducted by KEK.

1. Introduction

The High Energy Accelerator Research Organization (KEK) has been operating the J-PARC Muon Science Establishment (MUSE) since 2008. Among its four muon beam lines, the decay muon line (D-Line) has been extracting and providing positive and negative muon beam with a momentum range from several to 120 MeV/c (decay muon) as well as high intensity 30 MeV/c positive muon beam (surface muon) for a variety of science programs, utilizing a superconducting transport solenoid [1-3]. The D-Line as well as the other J-PARC facility suffered severe damages from the earth-quake on March 11, 2011. It necessitated rebuilding of the damaged superconducting transport solenoid. For this reason, a superconducting transport solenoid for the D-Line at J-PARC muon science facility was newly designed and manufactured. It generates a magnetic field in relatively large region (warm bore diameter 0.2 m), while keeping the same outer dimensions, connection interfaces to the existing

IOP Conf. Series: Journal of Physics: Conf. Series **897** (2017) 012022 doi:10.1088/1742-6596/897/1/012022

refrigerator and the power supply of the previous machine [1-3]. In addition, the solenoid provides stable operation utilizing conductive-cooling technologies as superconducting devices.

This paper describes the design, the manufacture of the solenoid, the magnetic field measurement at room temperature [4] and the results of performance tests conducted by KEK.

2. System Configuration

2.1. General Layout

Major specifications of the previous machine and the newly built solenoid are shown in Table 1. The solenoid consists of twelve pieces of 0.5 meter-long superconducting coils, which are series-connected mechanically and electronically. It is contained in a cylindrical vacuum vessel combined with an iron return yoke. All coils are indirectly cooled by cooling pipes with supercritical or two-phase helium flow supplied from the helium refrigerator. Thermal shields and current leads are also indirectly cooled by cooling pipes with gas helium flow. The major changes of both solenoids are the reduction of the rated current and the magnetic field, the equipment of a warm bore and the adoption of the high T_c current leads.

Table 1. Major Specifications of Solenoids.

Parameter	Previous machine [3]	Newly built solenoid
Magnetic Field	5.0 T	3.5 T
Diameter of Cold Bore	0.12 m	-
Diameter of Warm Bore	-	0.2 m
Length of Solenoid	6 m	6 m
Coil Shape	Solenoid	Solenoid
Wire of Coil	NbTi/Cu	NbTi/Cu
Number of Coils	12	12
Type of Current Leads	Gas cooled	high T_c

2.2. Magnetic Field

Figure 1 shows a schematic diagram of coils and iron yoke arrangement and magnetic field profile at the center axis of the warm bore. The entire solenoid generates 3.5 T along six-meter-long bore to confine and transport pions. Pions decay into muons during transportation in the solenoid due to their lifetime. Those muons are supplied to the downstream of the beam line. Since these coils are discrete, the magnetic field generated by the solenoid has the ripple of +/- 3 % maximum. The gaps are determined to be 20 mm, which is same value of the previous machine. However, according to the result of previous machine, the ripple is small enough to transport the muon beam without beam loss.

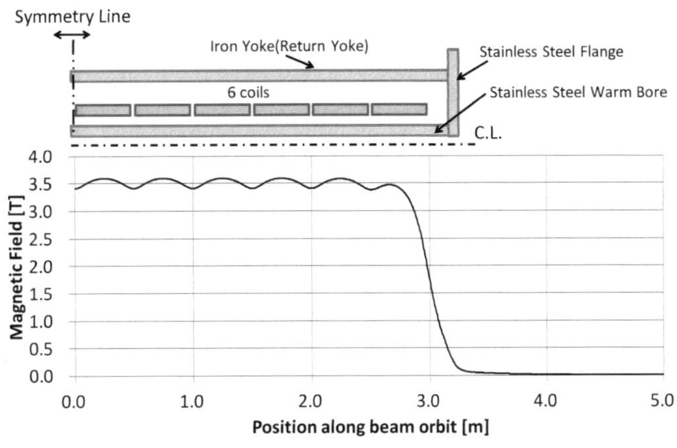

Figure 1. Schematic Diagram of the Solenoid Design and the Calculation Result of Magnetic Field.

IOP Conf. Series: Journal of Physics: Conf. Series **897** (2017) 012022 doi:10.1088/1742-6596/897/1/012022

3. Magnet Design

3.1. Coil and Iron Yoke Design

The design parameters of the previous machine and newly built solenoid, and the load line of the newly built magnet are shown in Table 2 and Figure 2.

The six-meter-long solenoid consists of twelve pieces of superconducting coils connected in series. Each coil was wound with approximately 3400 turns of NbTi/Cu superconducting wire. An aluminum alloy A5056 was adopted for the coil bobbins because the bobbins are the structural and the heat conductive member of the solenoid. The alloy has both relatively high mechanical strength and thermal conductivity in comparison with other non-magnetic materials. The coil dimensions at room temperature were determined to compensate thermal contraction (RT-4 K) of the constituent materials.

The iron yoke is a cylinder of outer diameter 0.6 m and has 30 mm thickness, which is required for magnetic flux return. The thickness was determined to reduce the leakage magnetic flux by the magnetic field calculation and to withstand a vacuum pressure and the transportation. Easily procurable low carbon steel (JIS SS400) was adopted for the iron yoke.

Approximately 40 km-long superconducting wires divided into 4 lots were used to manufacture the solenoid. The critical currents (I_c) of short samples which are collected from both ends of the lots were measured. When load line was calculated, we assume I_c as subtracting 3 sigma from its averaged value and define criteria for a load line ratio of less than approximately 75.0 % to avoid excess training quenches. The rated current is 415.7 A, corresponding to the load line ratio of 72.9 % to the short sample limit at 5.2 K.

Table 2. Parameters of Solenoids.

Parameter	Previous machine [3]	Newly built solenoid
Superconducting wire		
Type	Monolithic, NbTi/Cu	Monolithic, NbTi/Cu
Cu/NbTi ratio	3.89	4.8
Dimensions(insulated)	1.60 mm x 3.21 mm	ϕ1.56 mm
Magnet		
Coil inner / outer diameter	130 mm / 190 mm	300 mm / 330 mm
Number of Turns / Coil (design)	Approx. 3,000	3,377
Rated current	730 A	415.7 A
Central magnetic field / Peak field at conductor	5.0 T / -	3.5 T / 3.56 T
Total stored magnetic energy / High-field inductance	- / -	2.2 MJ / 25.5 H
Magnetomotive force / Coil (design)	2.2 MA	1.4 MA

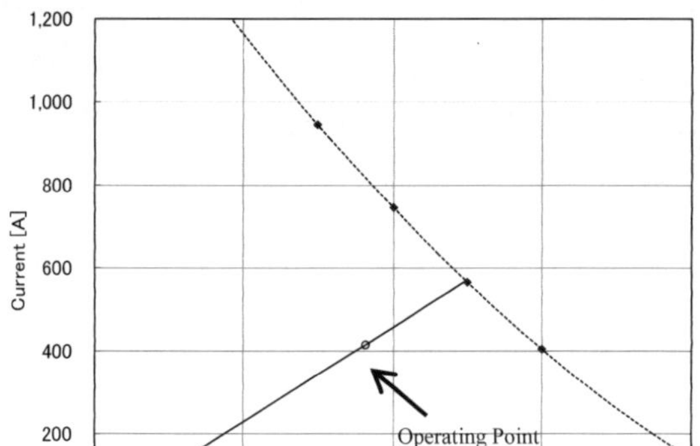

Figure 2. Load Line of the Magnet

IOP Conf. Series: Journal of Physics: Conf. Series **897** (2017) 012022 doi:10.1088/1742-6596/897/1/012022

3.2. Quench Protection

The electric circuit for the solenoid is shown in Figure 3. A sheet shaped quench-back heater was installed to the outer surface of each coil. The heaters are made of stainless steel, whose resistivity is not sensitive to temperature. After a normal zone emerges in one coil, the current bypassing starts when the current supply is shut down by detection of a quench. The generated heat causes quench in every coil, and the dissipation of magnetic energy in the entire solenoid. This avoids local heat concentration and enables to protect the solenoid.

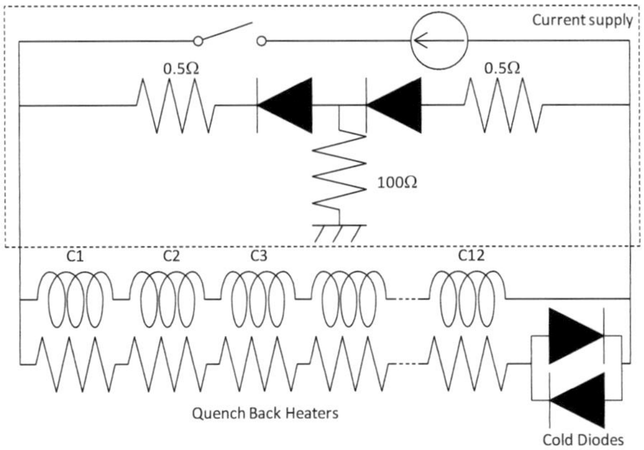

Figure 3. Electric Circuit for the Solenoid.

3.3. Cryostat Design

To reduce heat load of the cold mass, the aluminum thermal shields were installed to the inside and outside of the solenoid. A multi-layer insulation (MLI) was placed on the coil surface and the thermal shield surface. The total weight of the solenoid and the thermal shield is approximately 1,000 kg. CFRP cold-mass support system was introduced to bear this weight. In its current supply circuit, the high-T_c superconducting leads [5] were introduced between 4 K and 60 K in order to simplify the cooling scheme and the operation scheme of current leads in comparison with the previous machine. The summary of design heat loads is shown in Table 3. Both 4 K system and the shield system, design heat loads were much smaller than the refrigeration power. Therefore, stable operation was expected.

A chimney structure was attached to the downstream end of the solenoid, which serves as the interface to the refrigerator and the power supply. Non-magnetic stainless steel was adopted for the chimney vacuum vessel to avoid asymmetric magnetic field. In order to supply lower momentum muons compared to the previous machine, a warm bore with 0.2 m diameter was installed to the innermost area of the solenoid. A pillow seal flange was placed to the upstream end of the solenoid. A polyimide material was adopted to a part of MLI and instrumentation wires for its radiation resistance.

Table 3. Summary of Design Heat Loads

Source	Heat Load to Shield System [W]	Heat Load to 4K System [W]
Current Leads	43.2	0.5
Thermal Radiation	42.1	4.1
Magnet Supports	25.5	0.6
Instrumentation Wires	10.5	2.1
Transfer Tubes	4.3	3.1
Total	125.6	10.4
Refrigeration Power	200	35

4. Manufacture of Solenoid

Figure 4 shows an appearance of the completed solenoid for D-Line. In this picture, the near side is the upstream and the far side is the downstream of the beam line.

Figure 4. Completion of the Solenoid. An aluminum blind plate is temporarily flanged to protect the pillow seal flange face during the transportation (the near side of the figure).

5. Magnetic Field Measurement at Room Temperature

The magnetic field measurement using a Hall sensor along the beam axis was carried out after the completion of the solenoid. Very small current was supplied to the superconducting coils at room temperature. The value of the current was determined as 1 A. The magnetic field calculation results and the measurement are shown in Figure 4. The measurement had a good agreement with the calculation result in an error of less than +/- 0.5 %.

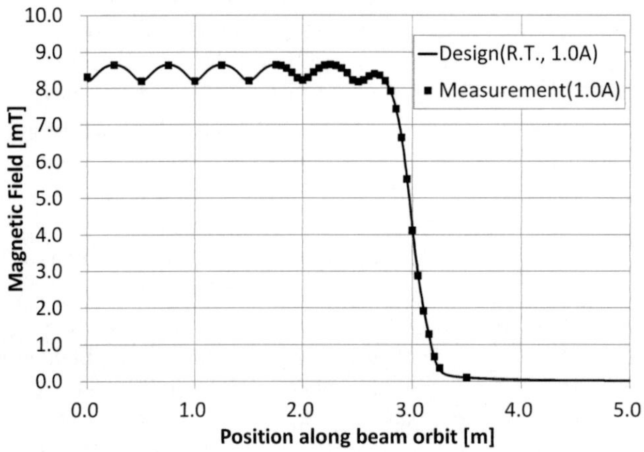

Figure 5. Magnetic Field Measurement at R.T. (Downstream) in Comparison with Calculation.

6. Performance Test

In order to confirm cryogenic and magnetic performance, the initial cooling test, the excitation test and the emergency shutdown test at the rated current were conducted by KEK.

6.1. Initial Cooling Test

The entire solenoid was cooled from room temperature over approximately 45 hours. The initial cooling curve is shown in Figure 6. The superconducting coils and thermal shields reached blow 4.5 K and 54 K, respectively. The cooling time shortened in comparison with the previous machine, whose cooling time is approximately 3 days [3].

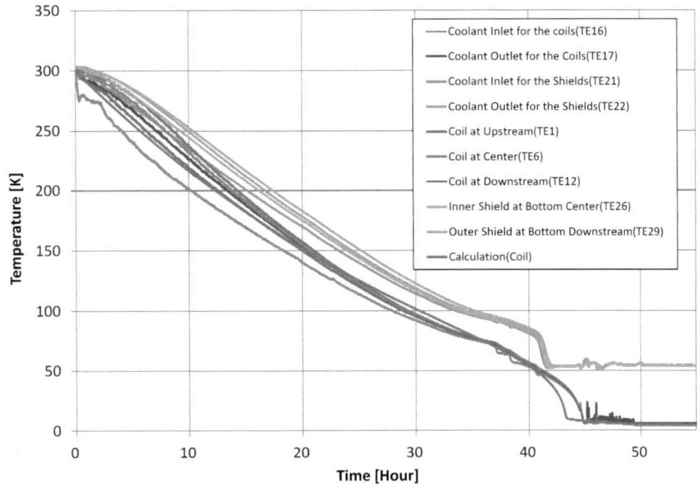

Figure 6. Initial Cooling Test from R.T (courtesy of KEK)

6.2. Emergency Shutdown Test

The emergency shutdown test at rated current was conducted after the solenoid reached the rated current without a spontaneous quench. The coil temperatures were measured by cernox sensors equipped to the surface of each coil bobbin. We simulated the temperature rise after a spontaneous quench in the adiabatic condition, which is the severe condition of the solenoid operation. The solenoid was designed to withstand the temperature rise, which is estimated as maximum 170 K.

The maximum temperature of the coils after the emergency shutdown test was approximately 51 K (Figure 7). It had an agreement with the design limitation of the temperature rise.

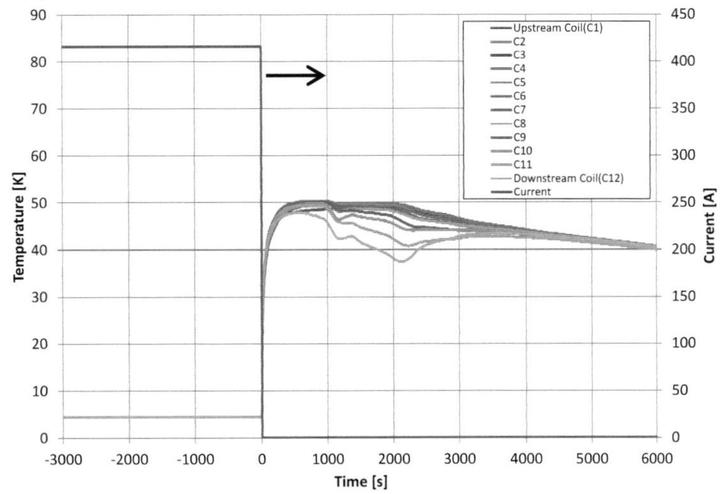

Figure 7. Emergency Shutdown Test (courtesy of KEK)

All tests were successfully performed with no damage. It is confirmed that the solenoid was precisely manufactured and fulfilled the requirements.

7. Conclusion

A new superconducting transport solenoid for D-Line of J-PARC muon science facility was designed and manufactured. It is a replacement for a severely damaged superconducting solenoid from the 3.11 earthquake in Japan. All tests were successfully performed with no damage. This result indicates that the solenoid was precisely manufactured and fulfilled the requirements. The solenoid was installed to D-Line in July, 2015 and has been under operation.

References

[1] K. Nagamine et al., “Superconducting Solenoid and its Cooling System for Pulsed Muon Channel” IEEE Trans. Mag., vol. MAG-17, no.5, p.1882, Sept. 1981.

[2] K. Nagamine, “Large Superconducting Solenoid and its Cooling System for Muon Channel” Teion Kou-gaku, vol. 20, no.4, p.187, 1985. (In Japanese)

[3] K. Shimomura et al., “Muon Beam Line at J-PARC MUSE -Reuse of a 30-year-old Superconducting Magnet- ” Teion Kougaku, vol. 45, no.4, p.174, 2010. (In Japanese)

[4] T. Semba et al., "Design and Manufacture of a Superconducting Solenoid for D-Line of J-PARC Muon Facility" , in Proc. 7th International Particle Accelerator Conference (IPAC'16), pp. 1177-1179, May 2016

[5] H. Teshima et al., “Recent Progress in Compact, Ro-bust and Superior Field-tolerant QMG Current Leads using RE-Ba-Cu- Bulk Superconductors” IEEE Trans. Appl. Supercond., vol. 26, no.3, 4800204, April. 2016.

Design Optimization for Superconducting Bending Magnets using Pareto Front Curve

Yukihiro Murata, Mitsushi Abe and Ryuya Ando

Hitachi, Ltd., Research and Development Group, 7-2-1 Omika-cho, Hitachi-shi, Ibaraki-ken, 319-1221 Japan

E-mail: yukihiro.murata.sf@hitachi.com

Abstract. A novel limit design method for superconducting magnets is presented. It is particularly suitable for ion core magnets such as those used in accelerator magnets. In general, a stochastic optimization whose objective functions consist of values, e.g., the magnetic field, experience field of superconducting coils, current density, and multipole field integral, is often used. However, it is well known that the obtained solution strongly depends on the initial one. Furthermore, once the calculation model is fixed, the range of solutions is also fixed, i.e., there are times when it may be impossible to find the global optimum solution even with a lot of parameter sweeps. In this study, we draw the Pareto front curve to obtain the range and infer whether the solution is an optimum one. In addition, the Pareto front curve indicates the neighborhood solution that is substituted for the initial one. After this process a stochastic optimization is implemented with its initial design parameters. To confirm the validity, we designed a superconducting bending magnet, and it showed that this method works well.

1. Introduction

Stochastic optimization methods have often been used in engineering design [1-3]. These methods have many merits such as easy handling. The optimization algorithm minimizes the single objective function or multiple objective functions. The single objective optimization requires breaking down several constraints into a single objective function while considering the differences in sensitivity of each function against the parameters. As a countermeasure, we have already proposed a method to generate an objective function with automated weighting [4]. In cases where a lot of constraints exist, the weighting is even more critical. In this technique, the ratio of the evaluated function value over each penalty function value is used to determine each weight. Each function value is smoothed through spline fitting. These weights should be updated automatically at predefined times. If they are constant the relationship among their function values changes with times. This automated weighting technique can lower the values of each function impartially, thereby avoiding obtaining local optimum solutions. The multiple objective optimization can estimate each constraint partially and obtain many non-inferior solutions. However, this needs a huge amount of calculation because there are a lot of design cases.

Furthermore, the optimization of design parameters is often conducted with the calculation model fixed. Once the model is fixed, the obtained solution could be limited to the area where the global optimum solutions does not exist, i.e., the global optimum solution may not be included in the solution set even with the use of an advanced variable optimization.

For a practical approach to dealing with optimization problems, the concept of Pareto optimization [5] has been proposed. A solution space has a boundary, called the Pareto front curve, and we can find a set of optimum solutions from the curve. The Pareto front solutions are not dominated by any other feasible solution, i.e., no other solution exists that is better for at least one objective function value and equal or superior with respect to the other objective function values. This fact has been widely used in engineering to aid designers in their decision-making processes. In this study, we proposed a method: at the first step determine the possibility of satisfying the demanded specifications from the Pareto front curve and then implement the optimization with the neighborhood solution as the initial solution. To draw the Pareto front curve we used one of the multi-objective optimizations, PSO (particle swarm optimization). PSO enables effective plotting on the solution space without wasteful computation. A feature of this flow leads to avoiding obtaining the local optimum solutions. The proposed method was applied to a bending magnet design, which showed good performance in solving this problem.

2. Design method

2.1. Drawing Pareto front curve and determining computational model

A solution space typically exists under inequality constraints, and its boundary is composed of a set of Pareto optimum solutions. In this study, PSO was used to generate the solution space. This design set of PSO parameters is shown in Table 1. The PSO concept was first introduced by Kennedy and Everhart in 1995 [6], and they showed that problems are optimized by having a population of candidate solutions, using dubbed particles, and moving these particles around in the search space according to simple mathematical formulae that describe the particle's position and velocity. As shown in Fig. 1, several sets of initial parameters generate the Pareto front curve through successive alternation of generations.

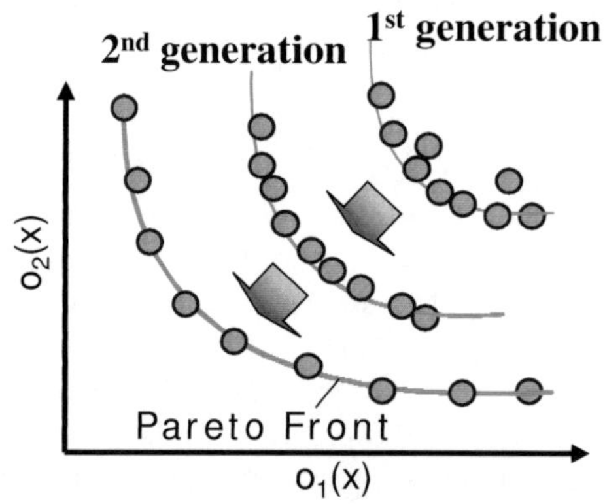

Figure 1. Conceptual diagram of Pareto front generation.

This curve indicates the boundary of the solution space, i.e., we can judge whether the current model is a solution candidate through comparative evaluation with the required specifications. The curve also gives the neighborhood solution. With this solution as the initial solution, we can optimize the parameters as described below.

Table 1. Design parameters of PSO.

Parameter	Value
Number in population	3000
Number in generation	25
Number in parameters	18
Maximum archive number	5000
Sharing	Crowding distance

2.2. Single objective stochastic optimization

In the previous section, the computational model was determined using the Pareto front curve. To optimize the rest of the parameters, an adaptive simulated annealing (ASA) method [7] was chosen as the single objective optimization algorithm. ASA departs from traditional simulated annealing (SA) by considering a generating probability density function with fatter tails than the typical Boltzmann distribution, allowing ASA to possibly escape local minimum solutions by considering farther points. The objective function should be created by summing each constraint function with weights. ASA has some characteristic features, for example, the parameter sensitivity, that is, the gradient of the objective function, can be estimated, separate annealing temperature parameters are assigned for each variable, and they are updated as the optimization progresses.

The entire computational flow in this study is shown in Fig. 2.

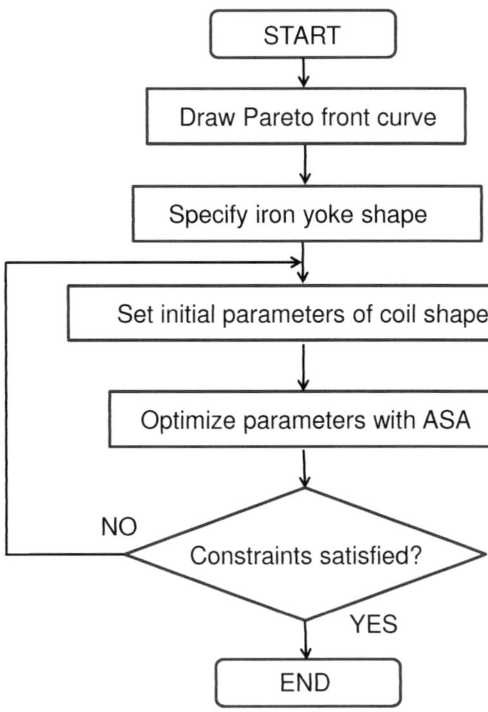

Figure 2. Optimization flow chart.

3. Optimization example

3.1. Modeling

A bending magnet design was implemented as an example of the proposed optimization. A schematic drawing of the model is shown in Fig. 3. This magnet is mainly composed of an iron yoke and a pair of superconducting race-track-type coils. The cross-sectional area of the superconducting coil was assumed to fit in a square of side 55 mm based on the result of the investigation of an optimized wiggler design for SAGA-LS [8]. The constraint list for this optimization is shown in Table 2. For an engineering design, a value close to the constraint, that is, a limit design is favorable. The magnetic field was calculated by using a commercial FEM solver (EM Solution, Science Solutions International Laboratory, Inc., Japan). The calculation points were set on the beam axis (z-axis) at 5 mm intervals for the dipole field integral, on the plane of y = 0 at 1 mm intervals along the z-axis and at 0.5mm intervals parallel to the x-axis for the sextupole field integral calculation. The field distribution around the beam axis is expressed in Eq. (1) [9].

$$B_y + iB_x = \sum_{n=1}^{\infty} (b_n + ia_n)(x + iy)^{n-1}. \quad (1)$$

To calculate the multipole field integral described later, we found the approximate curve to the distribution parallel to the x-axis as shown in Eq. (2).

$$B_y(x,0,z) = \sum_{n=1}^{\infty} \frac{b_n(z)}{(n-1)!} x^{n-1} = b_1 + b_2 x + \frac{1}{2} b_3 x^2 + \frac{1}{6} b_4 x^3 + \cdots. \quad (2)$$

Table 2. Constraint list.

Constraint	Value
Center field	5 T
Dipole field integral	0.84 Tm
Sextupole field integral	<30 T/m
Maximum coil experience field	<6.5 T
Current density	<200 A/mm^2

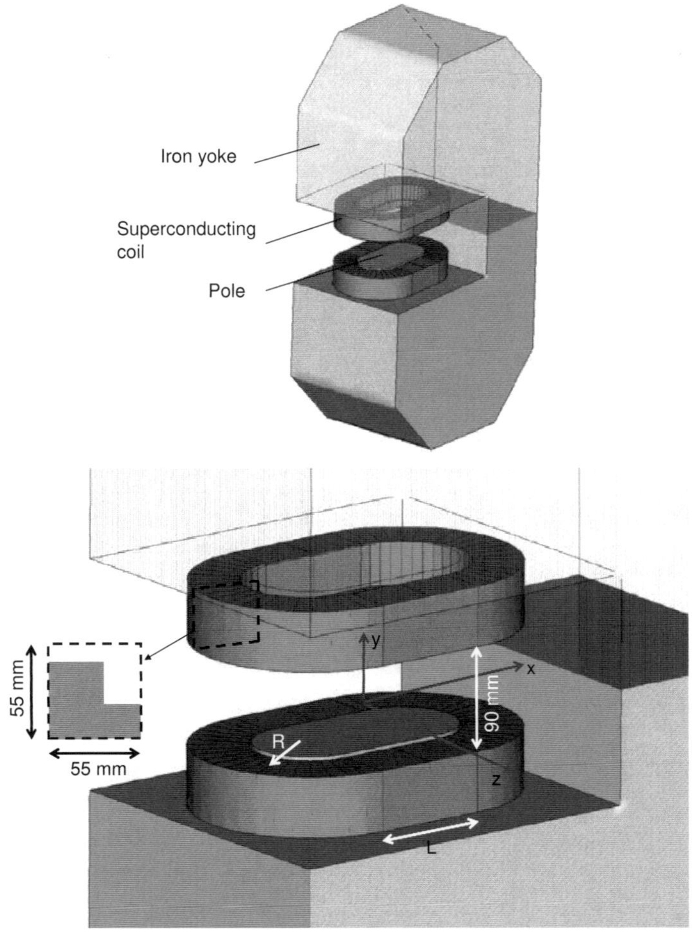

Figure 3. Schematic drawing of bending magnet with iron yoke. Race-track-type coil is arranged inside square of side 55 mm.

3.2. Drawing Pareto Front Curve

In the beginning, the maximum race-track straight length was assumed to be 126 mm. In consideration of an arbitrary coil cross-section, four coils were arranged in a single layer as shown on the left of Fig. 4. The current density was defined to be the same for each coil. The total number of design parameters is 18 (for 4 sets of coil center positions, each size of cross-section, the race-track straight length, and the current density). Figure 5 shows the results of plotting solutions. The horizontal axis is the sextupole field integral, and the vertical axis is the coil experience field. The circles mean with the race-track straight length of 126 mm and the squares mean with the other lengths. The solutions were plotted only when the value depending on the equality constraint was within a 2% error of the target value. This result shows that the Pareto front curve is made up of only the solutions of the 126 mm case, and a proper length exists in the condition where the sextupole field integral limit is 30 T/m. The sextupole field integral represents the spatial field homogeneity in the x direction, and the curved parts of the race-track affect the homogeneity, i.e., the sextupole field integral can be varied by changing the straight length of the race-track. The current distance from the beam axis (z-axis) to the curved parts is 126 mm/ 2 = 63 mm. As the magnetic field is spatially reduced at an exponential rate, the magnetic field gradient is represented as $\frac{\partial B_z(x)}{\partial x} = A\exp(-x/a)$ with constant values of A and

IOP Conf. Series: Journal of Physics: Conf. Series **897** (2017) 012023 doi:10.1088/1742-6596/897/1/012023

a. The sextupole field integral of the current model is approximately 13 T/m in the case of the length of 126 mm under the condition of the coil experience field of about 6.5 T (see Fig. 5). Then, the length corresponding to the sextupole field integral of 30 T/m was roughly estimated to be $\ln(30/13)\times 126$ mm = 106 mm.

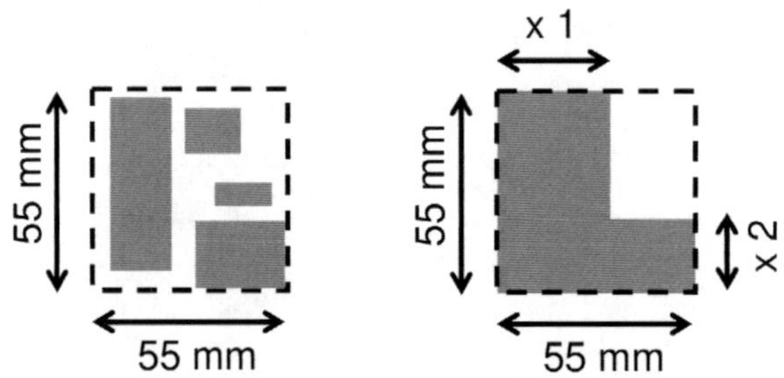

Figure 4. Coil cross-sectional diagram for drawing Pareto front curve (left) and for single objective stochastic optimization (right).

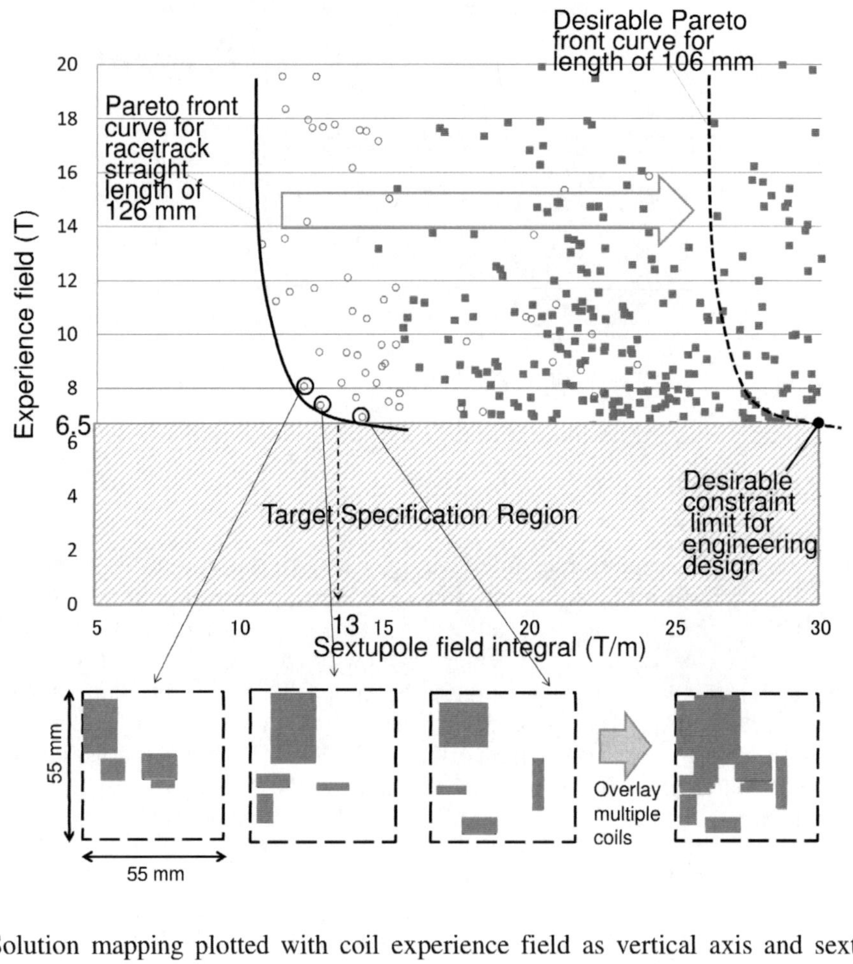

Figure 5. Solution mapping plotted with coil experience field as vertical axis and sextupole field integral as horizontal axis. Coil cross-sectional diagrams are shown with each Pareto solution, and superimposed drawing represents coil shape tendency. Circles mean solutions with race-track straight length of 126mm and squares mean with the other lengths.

3.3. Single objective stochastic optimization

Due to the above estimation, the race-track straight length was fixed. In the next step, the rest of the parameters regarding the coil cross-section shape and its current were optimized by using ASA. The single objective function $OF(\mathbf{x})$ was defined by the following equations.

$$OF(\mathbf{x}) = F_1(\mathbf{x}) + F_2(\mathbf{x}) + F_3(\mathbf{x}) + F_4(\mathbf{x}) + F_5(\mathbf{x}), \tag{3}$$

$$F_1(\mathbf{x}) = \left|(B_{center}(\mathbf{x}) - B_{center,target}) / B_{center,target}\right|, \tag{4}$$

$$F_2(\mathbf{x}) = \left|(BL(\mathbf{x}) - BL_{target}) / BL_{target}\right|, \tag{5}$$

$$F_3(\mathbf{x}) = \max(0, (G_S L(\mathbf{x}) - G_S L_{target}) / G_S L_{target}), \tag{6}$$

$$F_4(\mathbf{x}) = \max(0, (B_{\exp}(\mathbf{x}) - B_{\exp,target}) / B_{\exp,target}), \tag{7}$$

$$F_5(\mathbf{x}) = \max(0, (J(\mathbf{x}) - J_{target}) / J_{target}), \tag{8}$$

where $F_1(\mathbf{x}), F_2(\mathbf{x}), F_3(\mathbf{x}), F_4(\mathbf{x}), F_5(\mathbf{x})$ represent the peak field, dipole field integral, sextupole field integral, coil experience field, and current density, respectively. The vector $\mathbf{x}$ is a set of the design variables. The dipole field integral, $BL(\mathbf{x})$, is calculated with Eq. (9). This value is the integral calculated with magnetic field $b_1(z)$ (T) along the beam axis (z-axis).

$$BL(\mathbf{x}) = \int_{-\infty}^{\infty} b_1(z)dz. \tag{9}$$

The sextupole field integral, $G_S L(\mathbf{x})$, is an indicator of the magnetic field homogeneity. The sextupole field integral is represented with the coefficient $b_3(z)$ (T/m^2) as follows,

$$G_S L(\mathbf{x}) = \frac{1}{2}\int_{-\infty}^{\infty} b_3(z)dz. \tag{10}$$

At the bottom of Fig. 5, the coil cross-sectional diagrams that are neighborhood solutions are shown. The tendency of the four coil arrangements is that an empty region exists at the upper right corner, and the coil experience field becomes higher than 7 T, which means the current density is high. The spread of the cross-section area is an easy countermeasure to a decrease in the current density. In consideration of the above, the coil cross-section shape that was composed of two adjacent coils was defined as shown on the right of Fig. 4.

In the final step, an optimization with ASA was conducted. The transition of a single objective function against iteration number is shown in Fig. 6. After about 200 iterations, the feasible solution shown in Table 3 was obtained. The sextupole field integral is close to the target value of 30 T/m, which proves that the assumption of a race-track straight length of 106 mm was reasonable.

IOP Conf. Series: Journal of Physics: Conf. Series **897** (2017) 012023 doi:10.1088/1742-6596/897/1/012023

Table 3. Final constraint values through proposed method.

Constraint	Value
Center field	5 T
Dipole field integral	0.84 Tm
Sextupole field integral	28.7 T/m
Maximum coil experience field	6.3 T
Current density	184.8 A/mm^2

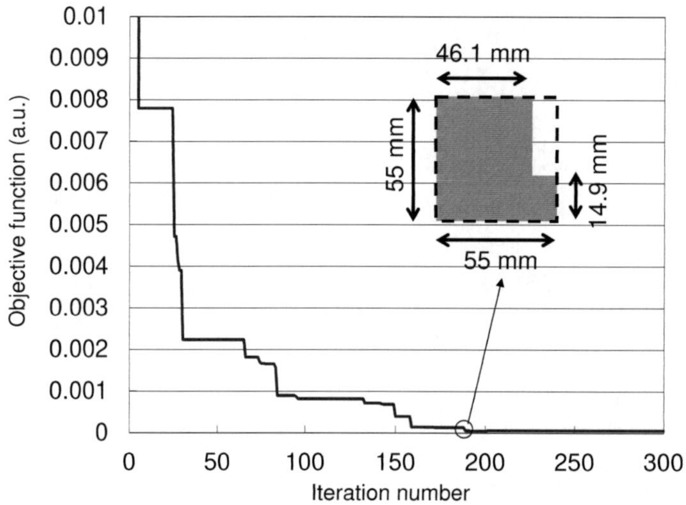

Figure 6. Transition of objective function against iteration number.

For validation, solutions with a race-track straight length of 66, 76, 86, 96, 106, 116, and 126 mm were plotted with Pareto front curves as shown in Fig. 7. The gradient of each curve is similar and is simply shifted along the horizontal axis. From Fig. 7, it can be seen that the sextupole field integral with the 106 mm model is about 28 T/m under an experience field of 6.5 T as expected, and the obtained solution is almost on the Pareto front curve. The modified Pareto front curve indicates which model could be an optimum candidate. Incidentally, regarding the experience field, the solutions over 10 T are apparently infeasible. However, these solutions are useful for drawing a Pareto front curve, because they are determined by factors related to not superconductivity but magnetomotive force.

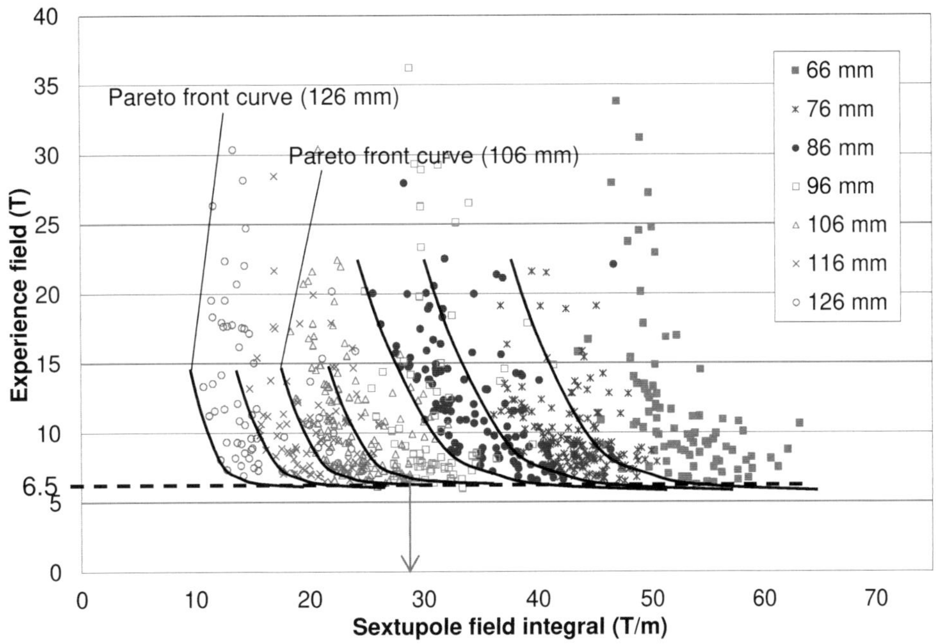

Figure 7. Solution plots with Pareto front curves as a parameter of race-track straight length.

The cross-section of this obtained coil was L-shaped. To confirm that this shape contributes to diminishing the sextupole field integral rather than the simple square shape, we drew the magnetic field distributions with both shapes. As shown in Fig. 8, the distribution with the L-shape is flatter than that with the square one, i.e., homogeneity was definitely improved.

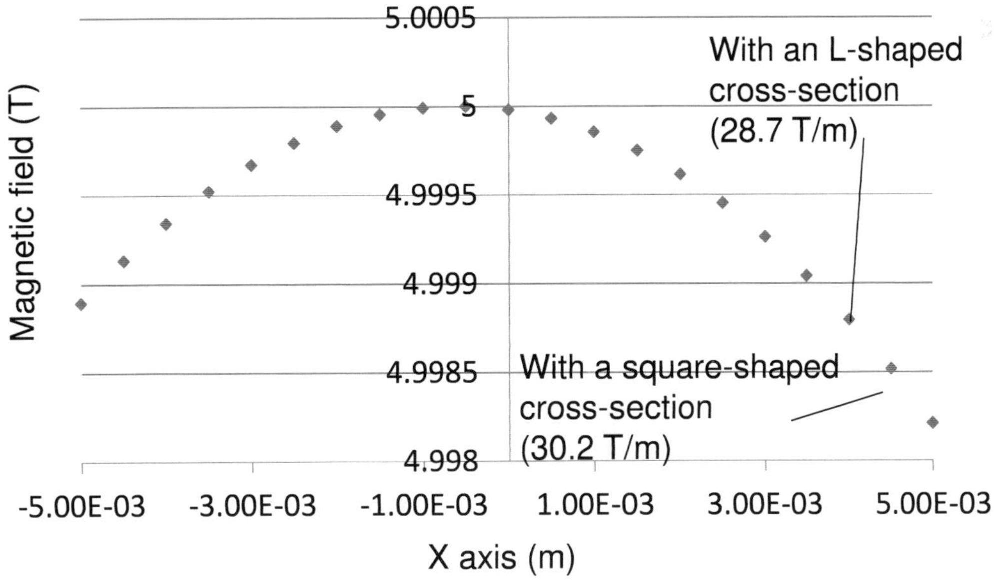

Figure 8. Magnetic field distribution along x-axis.

4. Conclusion

A new effective method for an optimal magnet design was proposed. A conventional optimization method has a higher possibility of falling into a local solution because of its high dependency on the initial parameters. The proposed method used a Pareto front curve for design parameter setting. The proposed method indicates the range of obtainable specifications with the current model. Based on this result, we could sophisticate the computational model and its initial parameters. It was shown that the optimization by the proposed method works well, and the solution reaches around the global solution as expected.

References

[1] T. Tadic and B. G. Fallone, "Design and Optimization of Superconducting MRI Magnet Systems With Magnetic Materials," *IEEE Transactions on Applied Superconductivity*, vol. 22, no. 2, pp. 4400107–4400107, Apr. 2012.

[2] Z. Ni, G. Hu, L. Li, G. Yu, Q. Wang and L. Yan, "Globally Optimal Algorithm for Design of 0.7 T Actively Shielded Whole-Body Open MRI Superconducting Magnet System," *IEEE Transactions on Applied Superconductivity*, vol. 23, no. 3, pp. 4401104–4401104, Jun. 2013.

[3] Y. Murata, M. Abe, R. Ando and T. Nakayama, "A Novel Design Method of Shapes of Ferromagnetic Materials for the Superconducting MRI Magnets," *IEEE Transactions on Applied Superconductivity*, vol. 19, no. 3, pp. 2293–2296, Jun. 2009.

[4] Y. Murata, R. Ando and M. Abe, "The Use of Automated Weighting to Generate an Objective Function by Penalty Method," *ISPJ Trans. Mathematical Modeling and Its Applications*, vol. 48, no. SIG19, pp. 99-106, Dec. 2007.

[5] J. A. Laponce, "Vilfredo Pareto, Manual of Political Economy. Trans. Ann S. Schwier, ed. Ann S. Schwier and Alfred N. Page. New York: Augustus M. Kelley, 1971, pp. xii, 504.," *Canadian Journal of Political Science*, vol. 5, no. 04, p. 599, Dec. 1972.

[6] J. Kennedy and R. Eberhart, "Particle swarm optimization," *Proceedings of IEEE International Conference on Neural Networks*, Piscataway, NJ. pp. 1942-1948, 1995.

[7] L. Ingber, Adaptive Simulated Annealing (ASA), [ftp.alumni.caltech.edu: /pub/ingber/ASA-shar,ASA-shar.Z, ASA.tar.Z, ASA.tar.gz, ASA.zip], Lester Ingber Research, McLean, VA, 1993.

[8] M. Abe, Y. Murata, T. Seki, T. Semba, Y. Iwasaki, Y. Takabayashi, T. Kaneyasu, and S. Koda, "Magnetic Field Design of a Superconducting Wiggler in the SAGA-LS Storage Ring," *IEEE Transactions on Applied Superconductivity*, vol. 24, no. 2, pp. 54–61, Apr. 2014.

[9] A. K. Jain, "BASIC THEORY OF MAGNETS," *Proceedings of CERN Accelerator School on Measurement and Alignment of Accelerator and Detector Magnets*, Anacapri, Italy, CERN-98-05, pp.1-26, 1997.

Magnetic Separation Using HTS Bulk Magnet for Cs-Bearing Fe precipitates

T Oka[1], K Ichiju, S Sasaki, J Ogawa, S Fukui, T Sato, M Ooizumi, K Yokoyama[2], S Aoki[3] and N Ohnishi[3]

[1]Niigata University, 8050 Ikarashi-Ninocho, Nishi-Ward, Niigata, 950-2181 Japan
[2]Ashikaga Institute of Technology, 268-1 Ohmae-cho, Ashikaga, Tochigi, 326-8558 Japan
[3]JNC Corporation, 2-1 Otemach, 2-Chome, Chiyoda-ku, Tokyo, 100-8105 Japan

okat@eng.niigata-u.ac.jp

Abstract. A peculiar magnetic separation technique has been examined in order to remove the Cs-bearing Fe precipitates formed of the waste ash from the withdrawn incinerator furnaces in Fukushima. The separation system was constructed in combination with high temperature superconducting bulk magnets which generates the intensive magnetic field over 2 T, which was activated by the pulsed field magnetization process. The separation experiment has been operated with use of the newly-built alternating channel type magnetic separating device, which followed the high-gradient magnetic separation technique. The magnetic stainless steel filters installed in the water channels are magnetized by the applied magnetic fields, and are capable of attracting the precipitates bearing the Fe compound and thin Cs contamination. The experimental results clearly exhibited the positive feasibility of HTS bulk magnets.

1. Introduction

In Fukushima, the combustion ash bearing radioactive Cs from disused incineration furnaces yields from the withdrawal process of incinerator furnaces. Then, it is necessity to recover Cs from the contaminated water extracted from the washed ash. The authors examined the feasible application of magnetic separation with use of HTS bulk magnets[1, 2]. The major merit of this magnet system is regarded as its intense magnetic field, high field gradient, and its compact system configuration[3, 4].

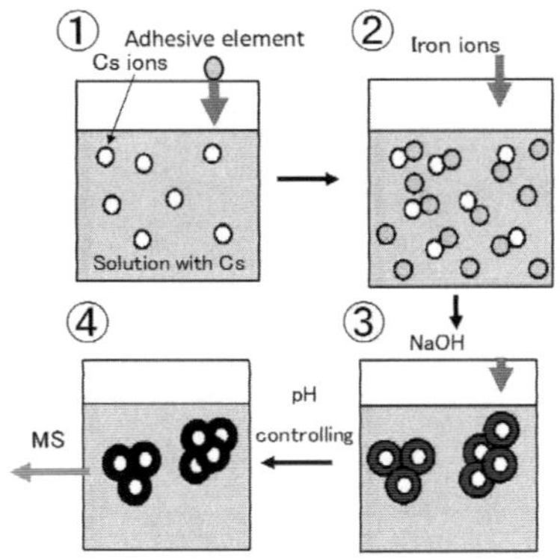

Fig. 1. Magnetizing procedure of Cs-bearing precipitates.

2. Experimental

2.1. *Precipitate forming*

Since Cs element, in general, is not 'magnetic,' it should be given the magnetic properties by combining ferromagnetic iron molecular. As shown in Fig. 1, first, the adsorbent was added

IOP Conf. Series: Journal of Physics: Conf. Series **897** (2017) 012024 doi:10.1088/1742-6596/897/1/012024

Fig. 2. Recovery of precipitate with use of conventional magnet roller.

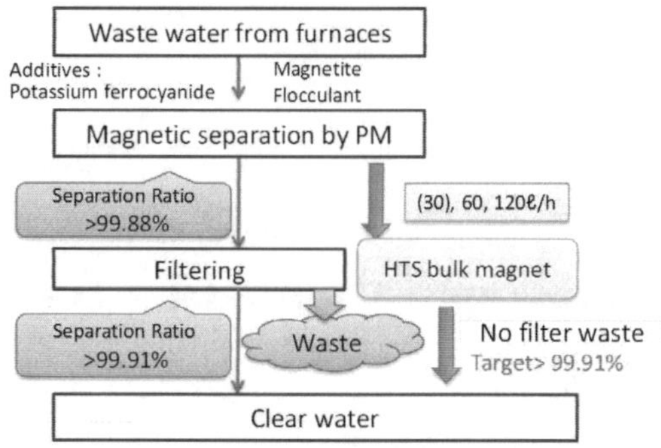

Fig. 3. Conventional purification process for Cs-bearing waste.

to the waste water, and then, the Fe-bearing particles were formed. The ferrocyanic compound was practically employed to form the precipitate after adding NaOH solution to control the pH value. The precipitate thus formed was roughly separated by the conventional magnet roller installing the permanent magnets. Figure 2 shows the photo of the separation process, showing the slurry attracted on the magnet roller.

The water thus purified was then led through the fine filter to collect the very fine precipitates remaining in the treated clear water. As the filter would possibly become radioactive by the substantially thin Cs solution, the purification process which releases no extra radioactive waste is preferable on this stage in order to purify the water to high level. The HTS bulk magnet is one of the candidates for this purpose, which needs no excess filters in the process.

As shown in Fig. 3, the sample water drained from incinerator furnaces is treated by the magnetic roller bearing 0.4 T permanent magnets, and the separation ratio reached over 99.88% at this stage. The flow rates in the magnet roller operations were chosen as 60 and 120 ℓ/h for the samples to which the flocculant was added beforehand to enlarge the particle sizes of the precipitate. In ordinary processes, the separation ratios reach over 99.91% by filtering the water in final, the target value of the separation ratio should be 99.91% when we employ the HTS bulk magnet.

The flocculant addition formed the coarse precipitates with the averaged sizes of 122.7 μm with concentration of 3.40ppm after the magnet roller process with the flow rate of 60 ℓ/h, and 149.6μm with 5.06ppm and 120 ℓ/h. In the case without any flocculant addition, the averaged size of the precipitates

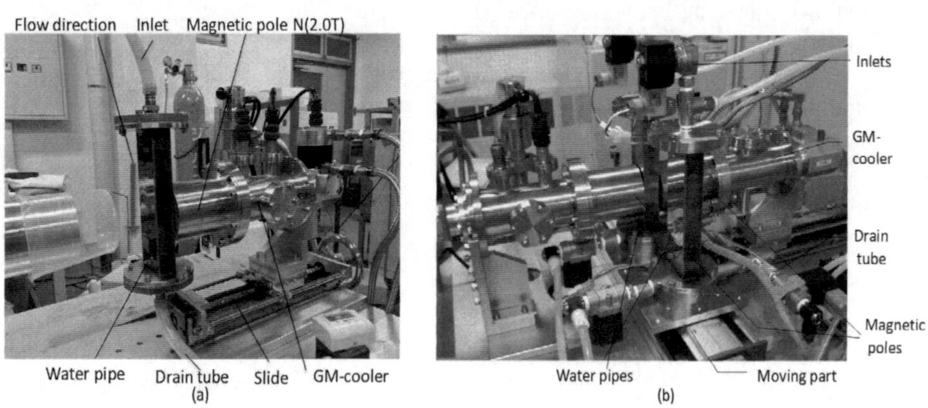

Fig. 4. Magnetic separation system carrying HTS bulk magnet system of mono-pole (a) and face-to-face poles with shifting channels (b).

was 4.10 μm with concentration 15.93ppm after the magnetic separation by the magnet roller in the slow flow rate of 40 ℓ/h.

2.2. Equipment for magnetic separation

Two types of the magnetic separation experiment were employed in the experiment. Figure 4 shows the systems using HTS bulk magnet of mono-pole (a) and face-to-face poles with shifting channels (b). The magnet was activated by the pulsed field magnetization method by feeding the current to the cryo-cooled pulse-coil, resultantly generating 2.03 T at the pole surface, which installs Gd-Ba-Cu-O bulk compound with the size of 60 mm in diameter, which was cooled to 30 K by GM-cooler[5, 6].

The water channel made of stainless steel (Japan Industrial Standard JIS:SUS304) with the size of 27 mm in thickness was attached on the magnetic pole, in which the stainless steel filter (JIS:SUS430) with the size of 0.1 mm in diameter was filled with the density of 10% in weight.

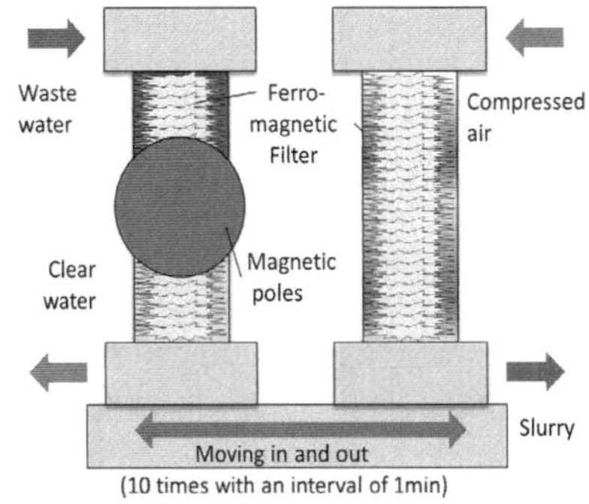

Fig. 5. Schematic illustration of HGMS with use of magnetic stainless filter and alternating water channels.

Another magnet system is a face-to-face type magnet system composed of face-to-face settled HTS bulk magnets. The components are the same as the former case. A couple of water pipes are positioned in the space between the 2-T magnetic face-to-face poles, which can move aside along the moving part in the figure. As schematically shown in Fig. 5, the pipes alternately move in and out of the intense magnetic field space to attract the slurry. So-called HGMS (high gradient magnetic separation) is operated in the pipe positioned between the magnetic poles, while the compressed air gushes in the water pipe positioned out of the field to clear the slurry away in every alternating 1 min interval. The flow rate of the water was 3.0 ℓ/min. The small amount of water samples were picked after each motion for ten times.

In the experiment, we employed the non-radioactive Cs fine powder. In exchange for thin Cs content, the concentration data for Fe ions were precisely estimated by ICP analysis by following equations, $S=(C-C')/C\times100$, where S means the separation ratio in %, C: concentration of original liquid, and C' : concentration of picked-up liquid. The flow rates were varied from 1 to 3 ℓ/min, as well as the separation time of 1 to 5 min.

3. Results and discussion

Figure 6 shows the separation ratios for the flocculant-added precipitate and without flocculant as a function of flow rates of 1-3 ℓ/min. When we add the flocculate, the performances reached over 90% for both cases of 60 and 120 ℓ/h samples, reaching the best separation ratios up to 96.80% and 99.89%,

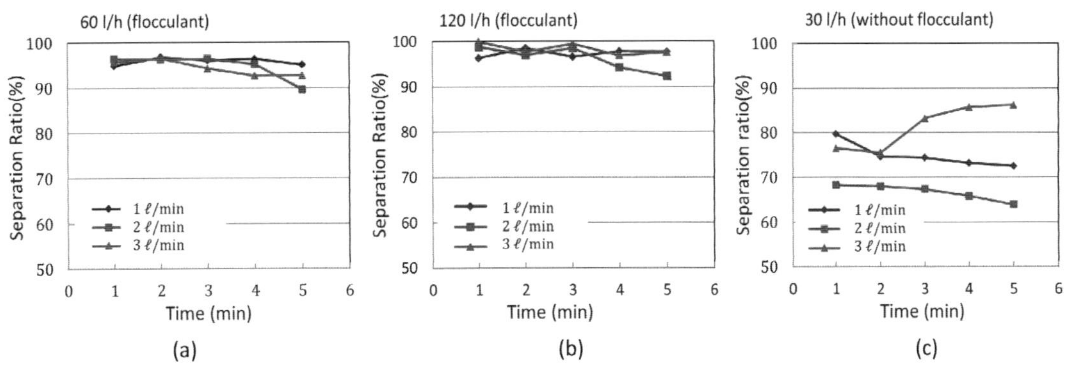

Fig. 6. Time dependence of separation ratios as a function of flow rates after one-way operation by mono-pole magnet for various flow rates in the magnet roller process.

respectively. As for the sample without flocculant addition, the performance stays in the region less than 86%. One sees that the flocculant addition, which means the enhancement in volume, is quite effective to obtain high separation performances, which follows the theoretical estimation[7]. Since we observed only weak descending against time in the figures, we see that the filter in the water channel has not reach its saturation, yet.

When we adopt the face-to-face type magnetic poles, the separation performance has a bit improved by its intense and wide magnetic field space between the poles. As shown in Fig. 7, the separation ratio data reached over 97% and 98% for 60 and 120 ℓ/h, respectively. The ratios for the sample without flocculant addition, however, stayed less than 67%, as well in the case of mono-pole operation. When we compare these data to the initial Fe concentration of 647.71ppm measured before the operation in the magnet roller process, the whole separation performance including the magnet roller process were estimated as 99.98% and 99.99% for 60 and 120 ℓ/h samples, respectively, as shown in Table 1. This performance exceeds the target value of 99.91%, exhibiting that the operation with use of HTS bulk magnet system is well profitable, as it was expected.

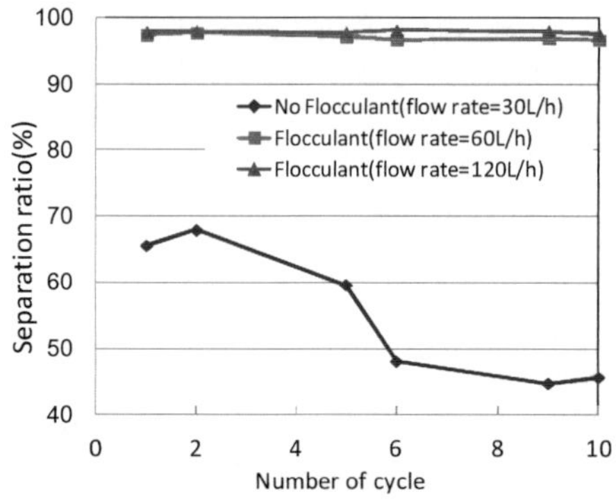

Fig. 7. Iterative cycle dependence of separation ratios as a function of flow rates by face-to-face magnets for various flow rates at the magnet roller process.

Table 1. Maximum separation ratios in the face-to-face HTS bulk magnet process.

Concentration(%)	60ℓ/h (with flocculant)	120ℓ/h (with flocculant)	30ℓ/h (without flocculant)
Before bulk magnet	4.82ppm	2.20ppm	9.24ppm
Final concentration (%)	0.11ppm(97.71%)	0.04ppm(98.11%)	2.96ppm(67.92%)
Separation ratio/initial	99.98%	99.99%	99.54%

Initial Fe-concentration: 647.71ppm, Flow rate: 3.0 ℓ/min

4. Conclusion

After the conventional processes by the magnet roller containing the permanent magnets which attract the major portion of precipitates, the separation performance of the bulk magnets reached up to 98 %, which exceeded the research target of 99.91%. Thus, the data suggest the feasible prospects for the high performance of magnetic separation for the purification of Cs-contaminated waste water. On the other, since the separation for the sample water without flocculant showed the low ratios, it is necessary to fabricate the coarse precipitate without adding any flocculant in order to reduce the whole mass of radioactive waste.

References

[1] Oka T, Kanayama H, Fukui S, Ogawa J, Sato T, Ooizumi M, Terasawa T, Itoh Y and Yabuno R 2008 *Physica C* **468** 2128
[2] Oka T, Kimura T, Mimura D, Fukazawa H, Fukui S, Ogawa J, Sato T, Ooizumi M, Yokoyama K, Tsujimura M and Terasawa T 2013 *Physica C* **484** 325
[3] Oka T 2007 *Physica C* **463–465** 7
[4] Durrell J, Dennis A, Jaroszynski J, Ainslie M, Palmer K, Shi Y-H, Campbell A, Hull J, Strasik M, Hellstrom E and Cardwell D 2014 *Supercond. Sci. Technol.* **27** 082001
[5] Oka T, Yokoyama K, Itoh Y, Yanagi Y, Yoshikawa M, Ikuta H, Mizutani U, Okada H and Noto K 2003 *Physica C* **392–396** 709
[6] Oka T, Yokoyama K, Itoh Y, Yanagi Y, Yoshikawa M, Ikuta H and Mizutani U 2000 *Physica C*

335 101
[7] Watson J 1973 *J. Appl. Phys.* 1973 **44** 4209